MIGRATION

Migration is in the news every day. Whether it be the plight of refugees fleeing Syria, or the outbreak of the Zika virus across Latin America, the modern world is fundamentally shaped by movement across borders. Arising from the 2018 Darwin College Lectures, *Migration* brings together eight leading scholars across the arts, humanities, and sciences to help tackle one of the most important topics of our time. What is migration? How has it changed the world? And what does it hold for the future? Authors approach these questions from a variety of perspectives, including history, politics, epidemiology, and art. Chapters related to policy, as well as those written by leading journalists and broadcasters, give perspective on how migration is understood in the media, and engage the public more widely. This interdisciplinary approach provides an original take on migration, providing new insights into the making of the modern world.

JOHANNES KNOLLE is an Associate Professor for the Theory of Quantum Matter at the Technical University of Munich. His work focuses on unconventional quantum phases of matter. He was awarded the Dissertation Prize in 2015 by the German Physical Society and the Springer Theses Prize in 2015.

JAMES POSKETT is an Assistant Professor in the History of Science and Technology at the University of Warwick. James's research engages broadly with the global history of science and technology from 1750 to the present day. His first book, *Materials of the Mind: Phrenology, Race, and the Global History of Science, 1815–1920*, was published in 2019. He writes regularly in the popular press, and in 2013 was shortlisted for the BBC New Generation Thinker Award.

THE DARWIN COLLEGE LECTURES

These essays are developed from the 2018 Darwin College Lecture Series. Now in their thirty-third year, these popular Cambridge talks take a single theme each year. Internationally distinguished scholars, skilled as popularisers, address the theme from the point of view of eight different arts and sciences disciplines.
Subjects covered in the series include

Migration

Edited by *Johannes Knolle*

Technical University of Munich

James Poskett

University of Warwick

CAMBRIDGE
UNIVERSITY PRESS

CAMBRIDGE
UNIVERSITY PRESS

University Printing House, Cambridge CB2 8BS, United Kingdom

One Liberty Plaza, 20th Floor, New York, NY 10006, USA

477 Williamstown Road, Port Melbourne, VIC 3207, Australia

314–321, 3rd Floor, Plot 3, Splendor Forum, Jasola District Centre, New Delhi – 110025, India

79 Anson Road, #06–04/06, Singapore 079906

Cambridge University Press is part of the University of Cambridge.

It furthers the University's mission by disseminating knowledge in the pursuit of
education, learning, and research at the highest international levels of excellence.

www.cambridge.org
Information on this title: www.cambridge.org/9781108746014
DOI: 10.1017/9781108778497

First published 2020

Printed in Singapore by Markono Print Media Pte Ltd

A catalogue record for this publication is available from the British Library.

Library of Congress Cataloging-in-Publication Data
Names: Knolle, Johannes, 1983- editor. | Poskett, James, editor. | Cambridge University Press.
Title: Migration / edited by Johannes Knolle, Technische Universität München, James Poskett,
    University of Warwick.
Other titles: Darwin College lectures.
Description: First Edition. | New York : Cambridge University Press, 2020. |
    Series: The Darwin College Lectures | Includes index.
Identifiers: LCCN 2019042622 (print) | LCCN 2019042623 (ebook) | ISBN 9781108746014
    (Paperback) | ISBN 9781108778497 (ePUB)
Subjects: LCSH: Emigration and immigration–Cross-cultural studies.
Classification: LCC JV6035 .M536 2020 (print) | LCC JV6035 (ebook) | DDC 304.8–dc23
LC record available at https://lccn.loc.gov/2019042622
LC ebook record available at https://lccn.loc.gov/2019042623

ISBN 978-1-108-74601-4 Paperback

Contents

Figures

Notes on Contributors

Fausto Bustos Carrillo is completing his final year of doctoral work in Epidemiology at the University of California, Berkeley. His primary advisor is Eva Harris. Fausto studies the clinical, epidemiological, and spatial aspects of Zika and chikungunya outbreaks in a Nicaraguan paediatric cohort. His research makes heavy use of infectious disease domain knowledge, epidemiological theory, semi-parametric biostatistical approaches, and the R statistical computing framework.

Iain Couzin is Director of the Max Planck Institute for Ornithology, Department of Collective Behaviour, and the Chair of Biodiversity and Collective Behaviour at the University of Konstanz, Germany. Previously he was a Full Professor in the Department of Ecology and Evolutionary Biology at Princeton University, and prior to that a Royal Society University Research Fellow in the Department of Zoology, University of Oxford, and a Junior Research Fellow in the Sciences at Balliol College, Oxford. His work aims to reveal the fundamental principles that underlie evolved collective behaviour, and consequently his research includes the study of a wide range of biological systems, from insect swarms to fish schools and primate groups. In recognition of his research he has been the recipient of a Searle Scholar Award in 2008, the Mohammed Dahleh Award in 2009, Popular Science's 'Brilliant 10' Award in 2010, the National Geographic Emerging Explorer Award in 2012, and the Scientific Medal of the Zoological Society of London in 2013. He coauthored one of the top five most cited papers in animal behaviour research of the decade 1999–2010.

Filippo Grandi became the 11th United Nations High Commissioner for Refugees on 1 January 2016. He was elected by the UN General

Assembly to serve a five-year term, until 31 December 2020. As High Commissioner, he heads one of the world's largest humanitarian organisations. The UNHCR has twice won the Nobel Peace Prize. Its 15,000-strong workforce spans 128 countries, providing protection and assistance to nearly 60 million refugees, returnees, internally displaced people, and stateless persons. Some 88% of UNHCR staff work in the field, often in difficult and dangerous duty stations.

Eva Harris is a Professor in the Division of Infectious Diseases and Vaccinology in the School of Public Health, Director of the Center for Global Public Health, and Chair of the Infectious Diseases and Immunity Graduate Group at the University of California, Berkeley. She has developed multidisciplinary approaches to study the virology, pathogenesis, immunology, epidemiology, clinical aspects, and control of dengue, Zika, and chikungunya, the most prevalent mosquito-borne viral diseases in humans. In 1997, she received a MacArthur Award for work over the previous 10 years developing programmes to build scientific capacity in developing countries to address public health and infectious disease issues. This enabled her to found a non-profit organisation in 1998, the Sustainable Sciences Institute, with offices in San Francisco, Nicaragua, and Egypt, which works to improve public health in developing countries by building local capacity for scientific research on priority diseases. She received a national recognition award from the Minister of Health of Nicaragua for her contribution to scientific development and was selected as a 'Global Leader for Tomorrow' by the World Economic Forum. She has published over 240 peer-reviewed articles, as well as a book on her international scientific work.

Johannes Knolle is an Associate Professor for the Theory of Quantum Matter at the Technical University of Munich. He was awarded his PhD in 2014 from the Max Planck Institute for the Physics of Complex Systems and the Technical University of Dresden. He held the Charles and Katherine Darwin Research Fellowship at Darwin College, Cambridge between 2015 and 2018. In 2017, he moved to Imperial College London as a Lecturer in Theoretical Condensed Matter Physics after a Marie Curie fellowship at the Cavendish Laboratory of the University of Cambridge. The focus of Johannes's work is on unconventional quantum phases of matter and materials. He received the Dissertation Prize 2015 of the Condensed Matter Section of the German Physical Society and the Springer Theses Prize 2015.

Chandran Kukathas is Dean and Lee Kong Chian Chair Professor of Political Science at the School of Social Sciences, Singapore Management University. Prior to this, he served as Chair of Political Theory at the London School of Economics and Political Science (LSE) between 2015 and 2019. Kukathas has also taught at the University of Utah, the University of New South Wales, Oxford University, and the Australian National University. He is the author of a number of books, including *Hayek and Modern Liberalism* and *The Liberal Archipelago*. He is currently working on a book entitled *Immigration and Freedom*.

David Olusoga is an Anglo-Nigerian historian, broadcaster, and author. Amongst other work, David presented and produced the BAFTA award-winning *Britain's Forgotten Slave Owners* for the BBC, in collaboration with UCL, and was the winner of the PEN Hessell-Tiltman prize for his most recent book, which accompanied the TV series of the same name, *Black and British: A Forgotten History*.

James Poskett is Assistant Professor in the History of Science and Technology at the University of Warwick. He received his PhD from the University of Cambridge in 2015 and held the Adrian Research Fellowship at Darwin College, Cambridge between 2015 and 2017. James's research engages broadly with the global history of science and technology from 1750 to the present day. His first book, *Materials of the Mind: Phrenology, Race, and the Global History of Science, 1815–1920* was published by the University of Chicago Press in 2019. He writes regularly in the popular press, and in 2013 was shortlisted for the BBC New Generation Thinker Award.

Kavita Puri is an award-winning broadcaster and executive producer. She presented the critically acclaimed Radio 4 series *Three Pounds in My Pocket* about post-war South Asian migration to Britain. For the 70th anniversary of the independence of India and Pakistan she presented the landmark Radio 4 series *Partition Voices*, which she is currently writing a book about. Kavita is also the Editor of the BBC foreign affairs programme *Our World*, for which she has executive-produced documentaries such as *Starving Yemen*, *The Killing of Farkhunda*, and *Freedom and Fear in Myanmar*. She was previously at *Newsnight*.

Venki Ramakrishnan received his bachelor's degree in physics from Baroda University in India in 1971 and his PhD in physics from Ohio University in 1976. He then studied biology for two years at the University of California, San Diego before beginning a postdoctoral position at Yale University. After a long career in the United States, he moved to England in 1999 to become a group leader at the MRC Laboratory of Molecular Biology in Cambridge. He is also the current president of the Royal Society. Ramakrishnan has a long-standing interest in ribosome structure and function, for which he won the Nobel Prize in Chemistry in 2009.

Khadija von Zinnenburg Carroll is an artist and the Professorial Chair of Global Art at the University of Birmingham. Khadija is the author of the book *Art in the Time of Colony* and the forthcoming *Bordered Lives*, based on her play *Shadows Talk* and the Immigration Detention Archive in Oxford. She is an editor of the journal *Third Text*. Her artworks about migration have been installed and performed at the Konzert Theater Bern, Pitt Rivers Museum, Pesta Bonka Festival, Indonesia, and Silver Sehnsucht, London.

Acknowledgements

We would like to thank all the different members of Darwin College, Cambridge, for their support in putting together both the lecture series and volume. This includes the Master, Mary Fowler, for hosting the events, as well as all the fellows and students who guided attendees in the packed lecture theatre on the day. Special thanks go to Torsten Krude and Andy Fabian for their initial enthusiasm for the idea, and for guiding us through the process. Also thanks are due to Espen Koht for his support with the IT and audio, and particularly for facilitating Khadija von Zinnenburg Carroll's performance lecture. Finally, we would like to extend a huge thank you to Janet Gibson for her exceptional support at every stage throughout the project. Without Janet, the Darwin Lectures just wouldn't be possible. Thank you again.

Introduction

JOHANNES KNOLLE AND JAMES POSKETT

Migration is in the news every day. Whether it be the plight of refugees fleeing Syria, or the outbreak of Zika virus across Latin America, the modern world is fundamentally shaped by movement across borders. This volume brings together eight leading scholars from the arts, humanities, and sciences to help tackle one of the most important topics of our time. What is migration? How has it changed the world? And what does it hold for the future? The present authors approach these questions from a variety of perspectives, including history, politics, epidemiology, and art.

The essays in this volume are based on a lecture series on 'Migration' organised by Darwin College, Cambridge, in early 2018. Naturally, the choice of topic and timing trigger associations with the exceptional surge of often forced migration to Europe from 2015 onwards. During the organisation of the lecture series, the United Kingdom voted to leave the European Union, further polarising public discourse on the subject. As a result, we now mostly think of migration as an exceptional event, something linked to political and humanitarian crisis. The motivation for a public lecture series on the topic was to redress the prevailing narrow debate surrounding migration. The contributions in this collection explore the many facets of migration, covering a wide range of topics, and getting away from a media focus on Europe and Brexit alone.

Traditionally, the Darwin Lecture series is a public multidisciplinary event bringing together renowned scholars and public figures from the arts, humanities, and sciences. It appeared to us an ideal setting for examining migration from different angles. We also wanted to take a global approach, moving beyond the so-called 'European migrant crisis', and instead exploring different aspects of migration in regions ranging from Latin America to South Asia. The hope was that there might be a common thread to migration as it appears in seemingly unrelated contexts. Indeed, the reader will find surprising parallels, for example, in how migratory birds and the spread of the Zika virus are influenced by globalisation; or in how new technologies are helping uncover the complexities of migration, both today and in the past. A number of contributors also emphasised the importance of addressing the legacies of empire and colonialism in order to understand migration today.

A common thread which emerges from the different chapters is that migration appears to be the norm rather than the exception. Migration is not an exceptional 'crisis' to be solved but rather an inevitable feature of our ever-evolving natural world as well as of human societies. That doesn't mean that migration is effortless, or that there aren't profound scientific and political questions posed by migration. Nonetheless, the starting point for any coherent approach towards migration cannot be a desire to erase it.

* * *

In the opening chapter, the historian and broadcaster David Olusoga examines the history of African migration to Britain. Olusoga argues that we need to understand this as a long history, dating back as far as the Roman period, if not before. This, in keeping with the theme of the volume as a whole, helps demonstrate that the presence of people of African descent in Britain is not exceptional, but rather a product of over two thousand years of history. Building on this argument, Olusoga highlights how many of the things people consider to be culturally 'British' are in fact the product of global migrations, often with roots in the world of slavery and empire. From drinking tea to eating fish and chips, the British are a people – like many others – profoundly shaped by a long history of migration.

Chandran Kukathas, Dean and Lee Kong Chian Chair Professor of Political Science at School of Social Sciences, at the Singapore Management University, then looks at the relationship between immigration and freedom. Typically, restrictions on immigration are seen as a restriction on the freedom of the migrant. However, Kukathas argues that in fact we should see restrictions on immigration as limiting the freedom of citizens in the host country as well. As Kukathas points out, much of the legislation on immigration restriction – both historically and today – is actually designed to restrict the freedoms of existing citizens.

Khadija von Zinnenburg Carroll's contribution is based on a special performance lecture exploring art and migration. Chair of Global Art History at the University of Birmingham, von Zinnenburg Carroll is also an active artist. The performance lecture and accompanying analysis explore the experience of immigrants in the United Kingdom's immigration detention centres. We have become used to migrants being figured as an abstract mass, and our legal system has evolved to keep them from public view in deportation facilities. Von Zinnenburg Carroll wants to take the migrant out of this shadow, both in her art and in her academic work.

The United Nations High Commissioner for Refugees, Filippo Grandi, then looks at the intersection of refugee movements and the more general phenomenon of human mobility today. He argues that migration is an intrinsic aspect of social and economic development, and that the current moment is not especially different from any other since the Second World War. Grandi also reminds us that Europe is actually only marginally affected by the arrival of refugees when compared with countries neighbouring those undergoing crisis. The world's poorer countries, such as Bangladesh, are often left to take in the majority of refugees, as in the recent case of Rohingya Muslims fleeing persecution in Burma. Grandi argues that the international response towards migration needs to be stable over the long term, in order to avoid exacerbating already difficult circumstances.

Kavita Puri, a British Broadcasting Corporation journalist, examines the history of migration during the Partition of India. Following the withdrawal of the British Empire in 1947, the subcontinent was split along religious lines into Hindu-majority India and Muslim-majority Pakistan. This resulted in one of the largest migration events in human history, with over ten million people moving between the two new countries. Many others, in the wake of the violence that followed, travelled to Britain, where they sought to rebuild their lives. Drawing on oral histories and interviews, Puri reconstructs the personal stories of Partition, examining this troubling history from different perspectives.

Eva Harris, Professor in the School of Public Health at the University of California, Berkeley, explores the migration of the dengue and Zika viruses across continents, around cities, and within the human host. In bringing together recent research results, Harris shows how viral evolution of diseases can be understood as the continued migration of mutations across time and space. Examples from Latin America highlight how a multidisciplinary approach across medicine, biology, and data science – combined with engagement of local communities – can overcome challenges to human health.

Venki Ramakrishnan, Nobel laureate and President of the Royal Society, then examines the relationship between migration and science. Ramakrishnan makes the case that science is a fundamentally international endeavour. Drawing on historical examples from the Middle East, India, Europe, and the United States, Ramakrishnan traces the way in which scientists have collaborated and migrated historically. He also reflects on his own career as a migrating scientist: born in India, studying and working in the United States, and now a group leader at the MRC Laboratory of Molecular Biology in Britain. Along the way, we learn about how Ramakrishnan conducted the research for which he was awarded the Nobel Prize in Chemistry in 2009, for the discovery of the structure of the ribosome.

Finally, Iain Couzin, Director of the Max Planck Institute for Ornithology, explores different examples of animal migration, from schools of fish in the ocean to insect swarms in the sky. Couzin shows how the coordinated response of a huge number of animals can emerge from surprisingly simple rules at the individual level. He also explores how the use of new technologies, such as drone tracking, as well as data science and even virtual reality, can help to reveal how organisms come together to face complex problems during migration across the globe.

1 Black and British Migration

DAVID OLUSOGA

I'd like to begin by asking you to think back to the 2012 Olympic Games. This was an event that I've come increasingly to see as the high-water mark of a phase in Britain's attitudes towards migration, integration, race, and identity. It was a vast, very expensive, very lavish – wonderful in my view – pageant directed by Danny Boyle. It was a celebration of British culture and British creativity. Like all Olympic ceremonies, it was designed to try to say something about the nation to whom the Olympic torch had just been passed.

Unlike most Olympic opening ceremonies, it was a resounding success. The history of Olympic opening ceremonies is a dark place of kitsch and cringeworthy moments. Danny Boyle created something remarkable; he succeeded where many others had failed in celebrating a national story. He celebrated our history and our culture as well as some of our most loved national institutions, above all the National Health Service. And the ceremony revelled in Britain's creativity and its pop culture. It celebrated our recent past, and it reminded the world of our unique place in youth culture, music, and creativity (Figure 1.1).

The Olympic opening ceremony also touched upon migration. Indeed, in celebrating the National Health Service it could not help but touch upon migration. After all, from its inception, the National Health Service has been dependent upon migrants, and remains so today. Among the moments from British history that were included in the opening ceremony that summer night in 2012 were the Industrial Revolution, the First World War, the struggle for women's suffrage, and the Jarrow March. Another event that was celebrated that night was the arrival in 1948 of the *Empire Windrush*. Indeed, a miniature replica of that ship was built, covered in printed replicas of British newspapers from the 1940s

FIGURE 1.1 National Health Service workers at the opening ceremony of the 2012 London Olympic Games. Credit: Alex Livesey/Staff/Getty Images Sport/Getty Images.

and 1950s. The replica ship was brought around the stadium by a team of twenty-first-century black Londoners, all dressed in the baggy suits, trilby hats, and flowery dresses of the 1940s. They looked exactly like the men and women who can be seen in photographs from that age; arriving in Britain, cold, disorientated, shocked by the temperature, carrying battered leather suitcases, and standing on the platforms of British railway stations. They looked exactly like their grandfathers and their grandmothers who'd arrived on the real *Windrush* 60 years earlier, and on the many, many ships that came afterwards.

At that moment in 2012, in an internationally televised moment, a pageant of national unity, an event millions watched and remembered, the arrival of the *Windrush*, the great symbolic event in the history of post-war migration and the black community, was set alongside other symbolic and great events in British history. After all, the arrival of the migrants in 1948, the year the *Windrush* docked at Tilbury, was also the year in which the National Health Service was founded. At that moment, the *Windrush* was ushered into the canon of British history: not just black British history, but British history.

Because I am culturally very British, although I am half Nigerian, I had, like many people, presumed that the opening ceremony would be a disaster. I'd feared it would be a national embarrassment. Instead I found myself, like many others I've spoken to, in tears. I've spoken to lots of black people for whom that moment of the *Windrush* being set alongside the Industrial Revolution, the First World War, and the Jarrow March was a moment of inclusion, a moment when they felt a sense of acceptance and belonging that momentarily felt like it was going to wipe away memories of racism and rejection.

Then, as the ceremony went on, there was more, there was the sight of young Londoners of all races and mixed races performing routines about youth culture. And there was a greater sense of belonging knowing that that generation will never know the racism of the *Windrush* generation. They'll never know signs that read 'no blacks, no Irish'. They'll never know what the slogan 'KBW' meant ('keep Britain white'), a slogan that Winston Churchill contemplated using as an official party slogan for the 1950 General Election until he was talked out of it by colleagues. And that younger generation would not know the racism I experienced in the 1980s, a decade when black people did not and could not trust the police, or believe that racial violence would be dealt with fairly or adequately. In the heart of the capital, with the world watching, Britain made a profound statement about the comfort and confidence that many people felt about migration and the place of the black community in our national story.

After the Olympics

Five years later, Britain to me feels like a very different place. Attitudes to migration feel very different; there is a gulf between then and now. Things are so different that you have to ask yourself how real was that moment in 2012: did it really happen and did the majority of people ever really think that way? Did we delude ourselves that we had become the nation that we told the world we were that evening in June 2012?

There were two things about the opening ceremony that those of us who celebrated it in the press missed. The first is that, being a celebration of Britain's creativity, it inevitably was a celebration of British youth. Inevitably, the performers were people from the demographic we know are far more comfortable than anybody else with globalism, integration, migration, and intermarriage. Secondly, and I think more importantly, the Olympic Games are awarded to cities, not nations. The 2012 Olympics were the London Olympics; they weren't the British Olympics. As a reflection of London, and of the attitudes of young Londoners, the ceremony was probably accurate. It caught the moment and who they were. What people learned since is that it did not capture the prevailing moods and attitudes of the country as a whole. It did not speak for the wider country.

What the 2012 Games did not reflect is a view that increasingly looks like one that has captured the imaginations and beliefs of a large proportion of this country. And it is a version of Britishness that also challenges British history, because there's a version of Britishness that requires, for its sustenance, a very particular version of the British past. In that version of British history, the Battle of Britain is a celebrated victory, but the Polish fighter pilots who contributed so much to that victory are forgotten. In that version of history, the Indian soldiers, who fought in both World Wars, are forgotten. This 'island story' version of British history is one that is resistant in multiple ways to stories of migration. It is also one that I fear is on the march. And yet if there's any nation on earth that can claim to have a culture and an identity that is pure, unadulterated, untouched by the outside world, unaffected by contact, migration, movement, and empire, that nation is not this one.

British Food and the British Empire

One of the best ways of appreciating how global the nature of our culture is – and how much of it is a result of migration, empire, and contact – is to take our national drink: tea. Tea, of course, originally came from China. The British liked it a lot, but they had nothing really to exchange it for that the Chinese wanted. Our main export was wool, which didn't enormously impress the Chinese. So the British began to grow the tea that they craved in India and Sri Lanka. (At the same time, the British went to war with China, first in 1839–42 and again in 1856–60, opening up the market and flooding it with opium.) By the end of the nineteenth century, there were over a million people in India and Sri Lanka growing tea for British consumption.

Unlike the Chinese, the British developed the taste for sweetening tea with cane sugar. Sugarcane comes from New Guinea, but the British encountered it first in Brazil. They borrowed the Portuguese technology for the cultivation and processing of cane sugar, which is one of the most difficult crops to grow, one of the most complex proto-industrial crops to process, and grew sugarcane on the islands of the West Indies. The plantations were worked by African slaves (Figure 1.2). This ultimate global drink was then consumed from porcelain teacups that were imported from China. So what we had by the beginning of the nineteenth century was an Asian crop grown on the islands of the Americas by Africans for consumption in Europe, where it was drunk with Chinese crockery. Historians have said that the only thing that's British in a cup of tea is the milk, but we imported our cows from Holland, so even that's questionable.

I'm not saying that what's great about the British cup of tea is that all of these things don't come from Britain. I'm saying tea is quintessentially, classically British because it is a global product, a product of movement, migration, and empire.

There's another example which I am fond of. Back in the noughties, there was a constantly repeated recycled news story centred on what was said to be our new national dish. It was reported that some poll or study, that may never really have existed, had suggested that Britain's favourite dish in the noughties was chicken tikka masala. The journalists who repeatedly recycled this story loved to point out that it was a form of cultural cuisine fusion as chicken tikka masala was a British–Indian hybrid dish. It originated in this country in the curry houses of northern England, not in the restaurants of Delhi or Kolkata, to which it was eventually exported. This, it was said, was an emblem of the nation that we had become: integrated, diverse, and comfortable with that. Many of these articles argued that it was a triumph of Britishness, that chicken tikka masala was a symbol of our great shift in British culture and history, an indicator of the fact that we had gone from the post-war age of separate communities to a more integrated Britain.

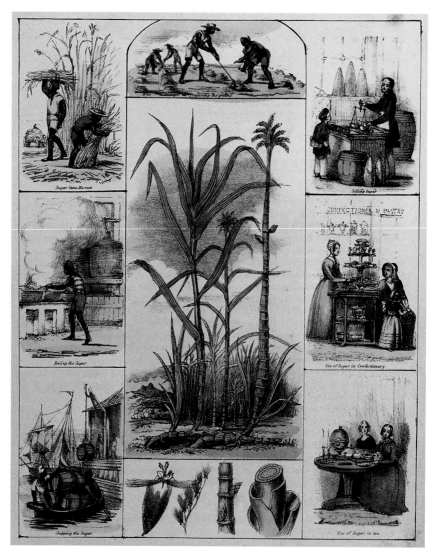

FIGURE 1.2 A lithographic print from the 1840s depicting a sugar plantation in the West Indies and the sale of sugar in Britain. Credit: Wellcome Collection, CC BY.

We'd accepted the migrants, the migrants had integrated, and our cultures had fused.

Many reports mentioned that chicken tikka masala had displaced fish and chips from the top spot as the nation's favourite dish. But the implication of those articles was that fish and chips was indigenous in a way that chicken tikka masala was not. Yet what was forgotten is that fish and chips was just as global, just as much a product of integration, migration, and movement as chicken tikka masala. Potatoes, of course, are not indigenous to the British Isles. They are part of that great Columbian exchange which means that virtually every day of our lives we eat foods that originated in the New World. Every day of our lives, most of us unknowingly say words from the Aztec language Nahuatl, words like 'tomato', 'potato', and 'chocolate'. As for battered fish, it was brought to Britain by Portuguese Marrano Jews, who arrived in the sixteenth century. The Portuguese brought the same innovation to Japan around the same time, where it was incorporated once again into another cuisine, and became a seemingly indigenous dish: tempura. Fish and chips, it turns out, is just as global as chicken tikka masala, but from an earlier forgotten age of migration and movement. The only difference is that this history has been forgotten, so much so that we convinced ourselves that fish and chips was invented here, that it was indigenously, organically British.

Migration in Roman Britain

There's a tendency among people like myself who think that migration is a good thing, that migration has helped us become the country that we are, to deny a reality that I think we need to acknowledge and accept: post-war migration was off the scale compared with earlier British migration. We do ourselves a disservice if we do not acknowledge that fact, and I believe we arm those who have a different view of migration when we do so. But those post-war waves of migration built on an older story. They were radically different in scale, but in nature they weren't so different.

However, historians who explore these older aspects of British history are under attack, as are the histories that they uncover. The producers of popular culture, television programmes, children's programmes, when they include storylines that depict non-white people in parts of British history before 1945, find themselves subject to waves of criticism and denunciation. This happened last year when the British Broadcasting Corporation (BBC) produced a cartoon in which the father of a central family set in ancient Rome was a centurion who was depicted as dark-skinned. This wasn't a new cartoon, it had just been picked up by people online using Twitter who had discovered it. But when it was discovered it became the focus of sustained online hostilities and attacks. The BBC was accused of having no respect for historical accuracy. It was accused of being influenced by political correctness and distorting the past for reasons of political correctness.

In the case of the BBC cartoon the historical facts are these: we know that there were people who lived in Roman Britain who would fit the modern definition of black, but we also know that the Romans didn't recognise race as we see it today. They didn't record it in their documents as a consequence. But geography, history, archaeology, and other sciences inform us of these facts. It is not that surprising when you think about it. At its height, the Roman Empire stretched from the coast of North Africa and Sub-Saharan Africa all the way to the north of England. The archaeological evidence is clear, and it has been augmented by new forensic techniques such as isotope analysis. Both the archaeology and the history indicate that Africans from both above and below the Sahara made their homes in the British Isles.

This research has given us figures like the Ivory Bangle Lady, a citizen of third-century York whose remains were discovered in 1901. In fact, the remains of the Ivory Bangle Lady were found in a stone sarcophagus alongside a number of ancient luxury items: blue glass beads, lockets made of silver and bronze, yellow glass earrings, and a beautiful blue glass perfume bottle. These objects all strongly suggested that this was the grave of a high-social-status resident of Roman York. Most significantly to me, she had two bracelets among her grave goods. One was made from ivory from Africa, where I was born, where my ancestors came from. The other bracelet was made of jet stone, which most likely came from Whitby on the northeast coast where my English family came from (Figure 1.3).

In 2009, 16 centuries after her death, the Ivory Bangle Lady was subjected to radioisotope analysis. This, combined with precise measurements of her skull and skeleton, and the chemical signatures left in her teeth by the food and drink she'd consumed, came together to show that

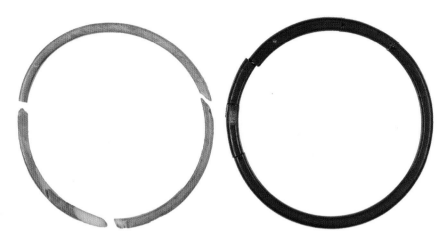

FIGURE 1.3 Bangles made from ivory and jet found in the grave of the Ivory Bangle Lady. Copyright Yorkshire Museum.

she was mixed-race but of North African descent. Her story and other similar research suggests that parts of third-century York could have been as ethnically and racially diverse as the city is today in the twenty-first century. To many people who go to the wonderful museums in York, this is just a fascinating revelation. It is also one of those historical facts that is interesting and surprising, and then, the more you think about it, becomes less surprising. The Roman Empire stretched across three continents. One of the things we associate most strongly with the Romans is roads: they liked to travel. We know that there were citizens from the Roman provinces of Britain in other parts of the Empire, so it is very unsurprising that the traffic was two-way, that those from Roman Africa came to Britain.

For most people this is an interesting novelty. It is a new fact that people welcome, just as we welcome learning new things about our country and our culture. But to some people, the existence of the Ivory Bangle Lady has been seen as a threat. None of this evidence – the science, the forensics, the history, the archaeology – can convince some people that what is taking place here is anything other than a conspiracy. Increasingly, there is a backlash against any attempt to demonstrate the presence of non-white people in Britain before 1945.

The Power and the Threat of History

I think it is important to keep these trends in perspective. To most people this is not a threatening history. It also is not entirely new. Historians have known about the black presence in Britain for a long time. Back in the 1990s the African-American historian Gretchen Gerzina was told by an assistant in a London bookshop that 'there were no black people in England before 1945'. She rather effectively disproved this by going on to write the classic book about black Georgians, of whom we know an enormous amount. Their faces can be seen in the portraits in the National Portrait Gallery and in collections all over the country. We have their diaries, letters, and books. In the 1720s, there was even a Jamaican man named Francis Williams studying at the University of Cambridge. Nonetheless, this weight of evidence is still something that some people are unable to accept.

The fundamental question therefore has to be, when so much evidence exists, when the weight of that evidence and the authority of the people who brought it to light is so unquestionable, why are some people threatened by this? Why are they offended by the idea of a black presence in Britain that stretches back before 1945? Why do they feel the need to dismiss science, history, and archaeology? Why are they at war with the facts? I think it is important to ask these questions because the refusal to accept that black presence in Britain has a long and deep history. It is a form of racism. I also think it is a rearguard attempt to defend an increasingly unsustainable monochrome vision of British history, a vision in which black people arrive for the first time on the *Windrush* and not before.

When I write about black British history before 1945, when I dig up details and explore lives, I'm accused of doing it for political reasons. That presumption that people like myself use history as a political tool is based on an assertion that we feel the need to defend our presence in Britain by exaggerating the duration and the scope of the historical presence of black people. I can only speak for myself here, and not for other scholars of black British history, but I am astonished that people feel that our sense of belonging and identity and citizenship is so fragile that we would need to distort history and fabricate evidence. And I think this accusation says an awful lot more about the people who make it than it does about the people it is aimed at. I would argue that it reveals a view that non-white immigrants can never be fully British.

When I watched the Olympic ceremony in 2012, I thought this idea had run its course and its day was over. It was expressed probably most toxically in 1968 by Enoch Powell in his infamous 'Rivers of Blood' speech. Powell claimed that a West Indian does not become, merely by virtue of being born in England, an Englishman. In law, he becomes a citizen of the United Kingdom, but by birth he is in fact still a West Indian or an Asian. What bothered Powell was that the generation of black Britons born in this country did see themselves as British, just as their fathers and their mothers who'd emigrated from the West Indies from Africa saw themselves as citizens of the British Empire. The idea that Britishness is a racial identity, or, perhaps more accurately, that Englishness is a racial identity, is a position that feels to me now more openly and regularly expressed than at any time since the 1980s, when it was shouted by skinheads on the streets of Britain to black people and their allies.

The history that counters that view, that shows black Britons as part of this country for centuries, is therefore itself contested. Historical evidence is being dismissed; research is being confronted with opinion or alternative facts. Two conflicting views of British history and British identity are beginning to emerge. Nowadays 2012 feels like a long, long time ago.

To me, the most clearly voiced expression of the form of British history that denies migration, that marginalises empire and its significance, was given in another speech by Enoch Powell, this time not so well known. It was a speech given on St George's Day in 1961. In that speech Powell explained to his audience that he believed that there was a deep providential difference 'between our Empire and those of others'. Powell said that what had made the British Empire so unique was 'that the nationhood of the mother country remained unaltered through it all, almost unconscious of the strange, fantastic structure built around her'. Britain, Powell said, was 'uninvolved'. In the same speech he went on to say that, 'when lucid connections which had linked her to distant continents and strange races fell away, it is our generation which comes home from years of distant wandering ... we discover affinities with earlier generations of English who felt no country but this to be their own, we discover affinities with them'. Powell then concluded with the following evocative and troubling image of the British past:

> Backwards travels our gaze beyond the Grenadiers and the philosophers of the eighteenth century, beyond the pipemen and the preachers of the seventeenth, back through the brash adventurous days of the first Elizabeth and the hard materialism of the Tudors and there at last we find them or seem to find them in many a village church beneath the tall tracery of a perpendicular east window and the coffering ceiling of a chantry chapel, from brass and stone, from line and effigy, their eyes look out at us, and we gaze into them, as if we would win some answer from their silence. Tell us what it is that binds us together; show us the clue that leads through a thousand years; whisper to us the secret of this charmed life of England.[1]

As Powell was making that speech, Queen Elizabeth was on her way to Sierra Leone for the ceremony that would mark the independence of Britain's oldest colony in Africa. Also, as he was making that speech, we were approaching the high-water mark of migration from the West Indies. The problem was that the 'strange races' hadn't fallen away. The strange races that Britain had been in contact with were people who had been partners and victims of five centuries of imperialism. Many of them had become British citizens. More of them aspired to become British citizens. Some of them at that moment were on their way with British Empire passports to these shores.

The very aspects of British history – empire, migration, cultural fusion, intermarriage – that Powell wanted to escape from and disavow were the aspects that explain why Britain looks the way it does today and why the great transformations that were happening in the post-war period were taking place. They are the stories that explain why people like myself – half English, half Nigerian – came to be in this country. These are the stories that explain how the Church of England arrived in Lagos, the

---

[1] Quoted in Heffer, S. (2014). *Like the Roman: The Life of Enoch Powell*. London: Faber & Faber, p. 335.

home of my father, an African city with a Portuguese name, as much a place of fusion as any other.

Powell's version of the British past is fundamentally a fantasy. The idea that we can commune with ancestors so distant, and in doing so discover an inner essence of Englishness, is a profound delusion. It is also a dangerous one. We are the peoples and the nation that was made by empire and migration, the people who eat fish and chips, and chicken tikka masala, and drink tea in Chinese cups. We are the result of 500 years of imperial history, five centuries of being a global power. There is no isolated, uninvolved, unadulterated England to which to return, even if we wanted to. Our nation and who we are simply doesn't make sense unless we understand the history of migration and the history of empire that drove much of that migration. Nobody put it better than the great academic intellectual Stuart Hall, who once said 'We are here because you were there.'

Further Reading

Collingham, L. (2017). *The Hungry Empire: How Britain's Quest for Food Shaped the Modern World*. London: Basic Books.

Eckardt, H. (2014). *Objects and Identities: Roman Britain and the North-Western Provinces*. Oxford: Oxford University Press.

Fryer, P. (1984). *Staying Power: The History of Black People in Britain*. London: Pluto Press.

Gerzina, G. (1995). *Black London: Life before Emancipation*. New Brunswick: Rutgers University Press.

Hall, S. (2017). *Familiar Stranger: A Life between Two Islands*. London: Allen Lane.

Olusoga, D. (2014). *The World's War: Forgotten Soldiers of Empire*. London: Head of Zeus.

(2016). *Black and British: A Forgotten History*. London: Macmillan.

Rappaport, E. (2017). *A Thirst for Empire: How Tea Shaped the Modern World*. Princeton: Princeton University Press.

Shyllon, F. (1977). *Black People in Britain, 1555–1833*. Oxford: Oxford University Press.

2 Immigration and Freedom

CHANDRAN KUKATHAS

Immigrant, n. An unenlightened person who thinks one country better than another

– Ambrose Bierce, *The Devil's Dictionary*

We wanted workers but got people instead

– Max Frisch

The prince who has more to fear from the people than from foreigners ought to build fortresses

– Machiavelli, *The Prince*

Living in Fear

Immigration is the subject that dominates politics in our time, most particularly in the liberal democratic Western countries of Europe, North America, and Australasia. While immigration has become an issue in many other nations – from Singapore looking to manage local attitudes to its large expatriate population, to Colombia facing an influx of desperate Venezuelans – in the liberal democracies it has provoked a backlash against outsiders as politicians as well as analysts of various stripes identify it as an existential question. Immigration for many is not just a source of unwelcome competition in the labour market or a problem for citizens struggling to gain access to over-subscribed public services from transport to health care. The threat it poses is not a financial one, or a matter of inconvenience, but an existential one.[1] Immigration poses a

---

[1] See, for example, Murray, D. (2017). *The Strange Death of Europe: Immigration, Identity, Islam.* London: Bloomsbury.

threat to fundamental liberal, democratic (American, Australian, Canadian, British, or European) *values*. It is for this reason above all that immigration must be controlled: to protect those values; and, by doing so, for these countries to take (back) control of their destinies. To some extent, it is the fear of losing control more broadly speaking that drives the imperative to control immigration.

The thesis I wish to defend here is that liberal democracies do indeed face a danger to the fundamental values that shape and define them, but the threat comes not from immigration itself but from immigration *control*. The search for control of the nation's destiny has been presented as a way of upholding its values and the means has been considered to be the securing of its borders from the impact of outsiders, whose movement needs to be more closely managed. This strategy is, however, deeply flawed, for the effort to control immigration cannot achieve its aim but nonetheless brings with it unintended consequences that undermine the aspiration that drives it. Immigration control in the end turns out to be less about controlling outsiders – or borders – than about controlling citizens. And the more vigorously that goal is pursued the more completely will control be exercised *not over outsiders but insiders*. The most troubling consequence of immigration control is therefore the loss of liberty and, over time, the transformation of society from one in which freedom is cherished and protected to one in which it is compromised and, ultimately, diminished.

To see this, we need first to understand what immigration is. The problem begins with questions of definition but extends further, for we need to come to terms with immigration as something that involves more than simply the movement of persons. It means seeing immigration as a practice that involves a particular form of behaviour which can be understood only in the context of a complex institutional structure. Once we better appreciate this, we will be able to grasp the nature of immigration *control*. After we have addressed problems of definition, we will be able to move directly to an explanation of why and how immigration control poses a threat to freedom, as well as to equality and the rule of law. Now, of course, it might be argued that even if these dangers are not negligible they should be discounted to the extent that there are important gains to be made by controlling immigration, either in the form of economic benefit or in terms of cultural protection. These arguments should be given careful consideration. It is also often argued that the most important gain to be made in the pursuit of immigration control is the recovery or re-assertion of a nation's right of self-determination. The later parts of our analysis will take a closer – and sceptical – look at these concerns before we return to our primary theme: immigration and freedom.

Definitions

What is an immigrant? Though this might seem like a straightforward question, it is not. This is because defining an immigrant is not a matter of trying to identify a natural kind. An immigrant is not a 'type' of person or an individual who falls objectively into a particular class or category. An immigrant is a person who is so classified because he or she has been determined to be so by the law's declaration or by the needs of organisations or institutions to group people into categories. The United Nations has thus defined an immigrant as someone who has been outside his or her country of nationality for one year or longer, and most social scientists use this definition when studying immigration and trying to count the number of immigrants in the world or in particular countries. Legal definitions of 'immigrant' are not always available: the United Kingdom, for example, does not recognise 'immigrant' as a legal category but distinguishes people according to whether or not they have a right of abode or residence in Britain.

The obvious question to ask here is why one year of residence should be the measure of whether or not a person is an immigrant. Why not six months or three months; or two years or five years? This is important because the number of people moving between countries makes it difficult to assume that mere border-crossing is sufficient to constitute immigration. Each year more than 370 million people cross the borders of the United States (including American citizens and residents) and 70 million cross the borders of the United Kingdom. Only a small fraction of these are considered to be immigrants, because visitors are not immigrants. But what makes someone a mere visitor? If it's just a matter of duration, then a grandmother visiting her family once who stays 367 days is an immigrant, while a seasonal worker who spends six months of every year in a country is not. Students who come to study for a 10-month Master's degree are not immigrants, but one who comes to earn a two- or three-year degree *is*. Thus, the British government has gone back and forth on the question of whether or not to include foreign students in the immigration statistics, since counting them pushes numbers up (and makes it look like there's an immigration crisis), whereas excluding them pushes figures down (and makes the government's claim to have reduced or controlled immigration more credible).

The other side of the question of what an immigrant is, however, turns out to be no less tricky. This is the matter of how to define a 'native' or a 'national'. If an immigrant is someone living outside his or her country of nationality, what is one's nationality? Are people with dual nationalities immigrants? Is nationality conferred by citizenship? Can an immigrant become a citizen and thereby become a national? What has to be appreciated is that the answers to these questions cannot be determined philosophically or theoretically, for there are many established definitions that have been used over time and hold in different parts of the world in a variety of legal and political systems. The law of nationality does not supply a fixed definition of the term but rather reveals a complex history of changing understandings of what it means to be a 'national'.

Consider the matter of nationality in the United Kingdom. The most authoritative work on the subject is Fransman's *British Nationality Law*, which runs to nearly 2,000 pages in its most recent edition. Much of it details the transformation of that law in the twentieth century, and a good portion of that recounts the changes that have taken place in the years after the war when Britain's colonies became independent. Take the example of Freddie Mercury, the lead singer of Queen, who was born Farrokh Bulsara in 1946 in Zanzibar (then a British protectorate) to Parsi parents who had moved there from India. At the time of his birth he held no citizenship but was simply a British subject until the British Nationality Act of 1948 created the status of 'Citizen of the United Kingdom and Colonies' (CUKC). There was no difference between the rights of CUKCs and other British subjects, and when the future rock star and his parents moved to Britain in 1964 it was as persons with the right to enter and live in the United Kingdom. Before the 1948 Act, all

British subjects in principle enjoyed freedom of movement and the right of abode within the Empire – though in practice they enjoyed only the right to move to Britain and not to other countries such as Australia and New Zealand, which imposed their own immigration controls. The earlier British Nationality and Status of Aliens Act of 1914 (which followed the 1911 Imperial Conference) had postulated that there was an equality of citizenship across the whole of the British Empire – without regard to race, ethnicity, or religion. This meant that at the Empire's height (around 1922) about 458 million people – a quarter of the world's population – shared a common nationality. But anxiety about the number of people from the Commonwealth moving to Britain led to the revision of British nationality law, which gradually restricted the freedom of British subjects outside Britain itself to move to the United Kingdom. The Immigration Act of 1971 effectively divided CUKCs into two groups, distinguishing those who did from those who did not have the right of abode in the United Kingdom. By 1983 six different tiers of nationality had been created under the British Nationality Act of 1981, and Commonwealth citizens ceased to be recognised as British subjects. Over the course of 50 years after the end of the Second World War, Britain lost an empire, and millions of people lost a nationality even as they acquired new ones.[2] Immigration control was not the only purpose that lay behind these changes, but it was undoubtedly the most important. The more general point of principle to be noted, however, is that immigration control is not straightforwardly about protecting the interests of nationals from would-be immigrants, for immigration control plays a critical role in the establishment of nationality.

The first step in the exercise of control is to determine who is to be subject to control, and in the case of immigration it means establishing

---

[2] In these few remarks I skip very quickly over an extremely complex subject. The most comprehensive treatment of the law of nationality here is Fransman, L. (2011). *British Nationality Law*, 3rd edn. London: Bloomsbury. For an excellent discussion of the subject, see Juss, S. S. (1994). *Immigration, Nationality and Citizenship*. London and New York: Mansell, especially Chapter 2, 'The growth of immigration controls in Britain', and pp. 51–3 on the British Nationality Act 1948. For American nationality law, see Aleinikoff, T., Martin, D., Motomura, H., and Fullerton, M. (2016). *Immigration and Nationality Laws of the United States: Selected Statues, Regulations and Forms*. Eagan: West Academic Publishing.

who is a native and who is foreign. One way of controlling the movement of people is by controlling through definition the number and type of people who need no permission to travel or to enter a territory or exercise a variety of rights that are denied to others. Immigration control begins not with walls or fences but with *classification*. That said, classification is not where control ends.

Immigration Control and Freedom

That immigration control begins with the classification of persons into insiders and outsiders is something that appears very obvious. But what bears emphasising is that very often classification involves turning people once thought to be insiders into outsiders. The history of British nationality law supplies only one such example of this process which gradually stripped people of rights they once possessed. American history is similarly replete with examples of legislation that revised and transformed the claims of people to be able to reside in, or be acknowledged as citizens of, the United States. This fact reveals something that needs to be more fully appreciated: the extent to which immigration control is as much about controlling citizens and residents – insiders – as it is about controlling would-be immigrants or outsiders.

The point of immigration controls is not simply to prevent entry into a state's territory or to limit the numbers who come in but to determine who may enter and to restrict what people who enter may do. Few countries wish to reduce the volume of cross-border traffic, if only because most want to encourage tourism or to attract business.[3] In 2013, 69.8 million people entered the United States as visitors, while more than 30 million entered the United Kingdom.[4] As we noted earlier,

---

[3] The US Department of Commerce records that the tourism industry in 2013 was valued at 2.8% of GDP, generating 8 million jobs and $1.5 trillion in travel and tourism sales. See http://travel.trade.gov/outreachpages/download_data_table/Fast_Facts_2013.pdf (accessed 3 November 2014), while the UK Tourism Alliance boasts that tourism's value to the UK could be priced at £1.26 billion or 9.6% of GDP (80% of which was non-domestic). See www.tourismalliance.com/downloads/TA_365_390.pdf (accessed 3 November 2014).

[4] Observing the increase of tourists coming to the United Kingdom in 2014, the Minister for Tourism said the figures were 'fantastic', even as the Government insisted that it was imperative that immigration to Britain was reduced. Minister

if citizens and residents are included, the numbers crossing American and British borders are even greater. While concerted efforts are indeed made to prevent people entering countries undetected by government authorities, the greater concern of governments is what those coming across the border do once 'inside'. The fear is that they will seek employment, or enrol in a school or university to study, or try to reside for an indeterminate period of time, or marry, or set up businesses, or engage in any number of otherwise legal activities.[5] The problem is that visitors arriving in such large numbers cannot easily be monitored, and if they seek to work or remain for longer than permitted there is little the authorities can do to keep track of their behaviour. This problem is exacerbated by the fact that citizens and residents are all too ready to cooperate with visitors by hiring them, teaching them, buying their wares, and generally helping them to stay – because they wish to take advantage of the cheaper or more skilled labour some visitors can provide, because they want to swell the ranks of their own groups with visitors of similar background, because they have something to sell, because they like their new-found friends, or because they fall in love. If visitors are to be kept from breaching the conditions of entry, it becomes necessary to monitor the behaviour of citizens and residents. For the restrictions under which visitors operate are largely restrictions on how they may cooperate with citizens and residents. If citizens and residents were disinclined to associate or cooperate with outsiders, the problem would never arise. Yet the propensity to truck and barter and to collaborate in various (questionable as well as innocent) ways is a deep feature of our nature, and foreigners will rarely find themselves welcome nowhere.

The only feasible way of monitoring and controlling would-be immigrants is to monitor and control the local population. It will be necessary to forbid among consenting adults not only capitalist acts, but also socialist, Christian, and more generally human ones – if one of those adults is a foreigner. Those in breach of laws forbidding such acts must be penalised – in the case of foreigners (and all too often, as we shall see,

---

for Tourism, Helen Grant, quoted in BBC News UK, 14 August 2014. See www
.bbc.co.uk/news/uk-28787769 (accessed 3 November 2014).
[5] Although there are often complaints that *immigrants* become involved in illegal activities or crime, this has not led to any reduction in entry targets for *visitors*.

citizens) usually by deportation and the denial of any liberty to re-enter the country in the future – or punished – by fines, the rescinding of rights and privileges, or imprisonment. Yet it is not just sins of commission that will attract the ire of the authorities. Citizens and residents will be expected to be careful to ensure that they do not cooperate or associate inadvertently with foreigners – and to keep records to demonstrate that commitment. Employers will have to monitor their employees, teachers their students, international carriers their passengers, parents their nannies, and doctors their patients. One nation under surveillance, its liberty diminished through unrelenting vigilance.

Now, it might be thought that this is a small price to pay for the many gains that immigration control brings. We will turn to the matter of the alleged gains in the analysis to come, but we should first look closely at the price, which is a larger one than many think. The loss of freedom in question is no trivial matter. We might begin by considering the fact that between 1930 and 2005, in its efforts to control and reduce immigration, the United States wrongly deported more than a million *of its own citizens.*[6] Wrongful deportation has not ended since the one million mark was passed, and the United States continues not only to deport hundreds of its own citizens every year but also to detain thousands of them who may take weeks or even years to prove that they are indeed Americans. The same is true in the United Kingdom, which in 2018 was found to have wrongly deported or detained thousands of its citizens and legal residents of West Indian descent who had arrived in Britain as a part of the *Windrush* generation – though the number of people harmed in this way remains unknown, just as it is still unclear how many people lost employment or were denied the right to rent accommodation because they were unable to prove their residency rights.

The price of immigration control is not one that is borne only, or even primarily, by immigrants. To understand this, it is necessary to see how it is that such injustices as wrongful detention and deportation come about, for it is not a bug in systems of immigration control but a feature. The trouble really starts with the anxiety to restrict the inflow of people

---

[6] See Motomura, H. (2014). *Immigration Outside the Law.* New York: Oxford University Press, p. 41.

as a result of the fear that outsiders will outcompete natives in the labour market by sheer numbers, or increase pressure on public as well as private goods and services (from health care to housing), or simply transform the social landscape by their very presence. This increases the pressure on government agencies and officials to find ways of excluding people and to ensure that those who enter the country do not enter the labour market, rent property, enrol in schools and universities, set up businesses, or engage in any of a number of activities that are entirely normal for citizens and residents.

On 30 July 2015, at the beginning of the 'Calais Crisis'[7] the British Prime Minister, David Cameron, promised to address the problem of large 'swarms' of people trying to enter the country by getting tougher on would-be immigrants.[8] Not only was his government ready to strengthen deportation measures but it would also take steps to make certain Britain did not become a 'safe haven' for people who had crossed the Mediterranean. To be sure, the government intended to do more to prevent people coming over the border to enter the United Kingdom; but it also aimed through such measures as deportation to strengthen what the Prime Minister called Britain's 'internal borders'.

What is noteworthy here is the recognition that borders are not to be found at political boundaries or in geographical space, but in laws and practical measures to control how people move and what people do. The Prime Minister was reported as saying that 'Britain's "internal borders" of a crackdown on housing entitlements and more removals' would make Britain less enticing to immigrants. Indeed, he went on to indicate that there was a need for further domestic reforms to deter immigration:

---

[7] In 1999 the opening of the Sangatte Refugee Centre in Calais attracted a large number of asylum seekers, prompting the French Interior Ministry to close the centre over the course of 2001 and 2002. Nonetheless, would-be immigrants and asylum-seekers continued to arrive and set up makeshift camps. Though there are no satisfactory figures for the numbers involved, a large proportion of the inhabitants arrived with the intention of crossing the channel into Britain. For an early account and critical analysis see Schuster, L. (2003). 'Asylum seekers: Sangatte and the Tunnel'. *Parliamentary Affairs* (Special Issue on Crisis Management), 56(3), 506–22.

[8] Reported by Holehouse, M. (2015), 'David Cameron blames Calais crisis on "swarm" of migrants', *The Telegraph*, 30 July 2015, www.telegraph.co.uk/news/uknews/immigration/11772410/David-Cameron-blames-Calais-crisis-on-swarm-of-migrants.html (accessed 13 August 2015).

'Frankly we also have to do more, and we are already passing legislation to do more, to make Britain a less easy place for illegal immigrants to stay.' In a particularly illuminating passage, he said 'Since I have become Prime Minister we have made it harder to get a driving licence, to get a bank account, to get a council house. We have removed more people. All of these actions – the internal border – matters, as it were, as well as the external border.' Given how difficult it is to distinguish legal from illegal immigrants, and immigrants from residents and citizens, the obvious question to ask is whether one can make life harder for illegal immigrants without also making it harder for everyone else. If the government makes it harder for 'illegal' immigrants 'to have a home, to get a car, to have a job, to get a bank account',[9] can it avoid making it harder for everyone to get these things? The answer to that question is plainly 'no'.

But there are other ways of making life more difficult for immigrants that also mean making life harder for citizens and residents that are more invidious still. There are several ways in which immigration law limits the freedom of the citizens and residents of a society. First, it limits their freedom to associate, whether for economic, cultural, political, or simply personal reasons: employers cannot hire whoever, or as many as, they prefer; universities often cannot recruit the people they most want; and individuals and families cannot come together as they wish. Second, it limits people's autonomy of judgment, as they are compelled to yield to the determinations of the authorities (or the democratic majority?) as to whether the people they hire, the students they admit, or the companions they choose are necessary or suitable for their own purposes. Third, it imposes not only compliance costs but also the risks that come with failure to comply – fines, loss of income, business collapse, institutional closure, separation from loved ones, or imprisonment.

Let us consider some examples of such forms of control by looking at the regulation of employers, educational institutions, and citizens more broadly. In the case of employers, immigration law imposes a substantial

---

[9] David Cameron, quoted in Jones, H. (2014). 'Are we going to be allowed to stay here? How government anti-immigration communications are infiltrating everyday life of British citizens', posted 14 July 2014, http://mappingimmigration controversy.com/2014/07/30/arewegoingtobeallowedtostayhere/ (accessed 15 August 2015).

regulatory burden – and frequently one that is too high for smaller businesses to bear, forcing them to yield an advantage to larger enterprises and corporations. Take the case of regulations brought in by Australia's Migration Amendment (Reform of Employer Sanctions) Act 2013, which was promulgated in September that year to revise the Migration Act 1958. It introduced new non-fault civil penalties for employers, adjusted existing criminal penalties, and expanded the liability of executive officers involved in employing 'illegal workers', while also establishing new investigative powers available to the Department of Immigration and Citizenship. Under the Act, Australian employers wishing to avoid sanctions must be 'proactive in checking the immigration status and work rights of all workers'.[10] Penalties begin at A\$15,300 for an individual and five times that for a corporate body, and apply regardless of the state of mind of the employer – that is, the Act makes no distinction between an employer who innocently allows someone to work without authorisation and one who knowingly disregards the law.[11] An employer who knowingly employs a foreign national to work illegally is also criminally liable to punishment by two years' imprisonment – or longer if there is evidence of worker exploitation, especially if it causes persons to enter into slavery, forced labour, or debt bondage. A person who operates a service, whether for reward or otherwise, referring other persons for work would be guilty of a civil offence if a referred person is a foreign national not legally entitled to work, or is recommended for work outside the terms of his or her visa.[12] The Act also established new powers to search premises, operate electronic equipment, download materials, seize evidence, and ask questions. The purpose of the Act was to address the problem posed by illegal workers, who were estimated by the Howells Report to number something less than 1% of the Australian

---

[10] Kinslor, J. (2013). 'New employer sanctions under the Migration Act'. *Law Society J.*, May, 34–5, p. 34.

[11] Employers might escape civil liability if they are able to establish that they held a reasonable belief that an employee was entitled to work – perhaps because the employee supplied fraudulent documents on application for a job.

[12] This would mean that not only commercial employment services but also community networks or student employment offices could be found liable for foreign workers' breaching of employment law.

workforce.[13] The employer sanctions specified in this Australian Act are not significantly different in character from those provided for in legislation by the European Union, or by the United States and Canada.[14]

While it is important to ensure that employers are not facilitating the activities of slavers and people traffickers, it has to be noted that the pressure on employers not to hire 'illegal' labour imposes a heavy burden. Settling the question of who exactly is ineligible to work in a country is a matter that is legally fraught and politically determined. It is not the case that there are obvious categories of people who must not be allowed to work. Employers want workers and search for them nationally and internationally. Governments permit them to do so and allow foreign workers into the country. But they allow some and not others. Who they admit turns less on the needs of the national economy than it does on the wishes of the more powerful and influential employers who are better able to secure exemptions or favourable legislation – though even they will face difficulties. In one way, of course, it is hardly surprising that the movement of labour is influenced by the wishes of employers: to whom else might governments turn when trying to determine which workers to admit and which ones to refuse? Adam Smith put the matter very clearly:

> The patrimony of a poor man lies in the strength and dexterity of his hands; and to hinder him from employing this strength and dexterity in what manner he thinks proper without injury to his neighbour is a plain violation of this most sacred property. It is *a manifest encroachment upon the just liberty both of the workman and of those who might be disposed to employ him.* As it hinders the one from working at what he thinks proper, so it hinders the others from employing whom they think proper. To judge whether he is fit to be employed may surely be trusted to the discretion of

---

[13] The Howells Report precipitated the amendment. See Howells, S. (2011). 'Report of the 2010 Review of the Migration Amendment (Employer Sanctions) Act 2007', Commonwealth of Australia, p. 25.

[14] See, for example, the 'Ad-hoc query on penalties and sanctions for employing illegal workers', requested by the United Kingdom European Migration Network on 20 January 2015, compiled 5 March 2014, http://ec.europa.eu/dgs/home-affairs/what-we-do/networks/european_migration_network/reports/docs/ad-hoc-queries/illegal-immigration/530_emn_ahq_penalties_for_employing_illegal_workers_05march2014_wider_dissemination.pdf (accessed 15 August 2015).

the employers whose interest it so much concerns. *The affected anxiety of the law-giver lest they should employ an improper person is evidently as impertinent as it is oppressive.*[15]

Employers are not, however, the only sector of society on whom the immigration law has placed a significant burden. Other parts of society have also felt its impact. One sector that is worth a closer look is the institutions in society responsible for education. To some extent, schools and universities face similar problems to those that affect employers, since they are employers themselves. But there are other issues that arise out of the distinctive purposes that educational institutions might have. A part of this is because schools and universities not only recruit international faculty to teach and administer, but also admit students from different parts of the world. By 2015 there were more than five million foreign students in higher education worldwide. In the United States, by 2009, higher education was the country's fifth-largest service export sector, with in-bound foreign students contributing $17.7 billion to GDP.[16] But it is not only the colleges and universities that are welcoming international students. In 2013 about 73,000 foreign students enrolled in American high schools. By 2015 more than 450,000 American students – predominantly but not exclusively college students – were themselves studying abroad, encouraged to do so by state education authorities and schools as well as by universities. The world has also seen a dramatic increase in the number of international schools catering primarily to the needs of 'expatriates', but also to local populations interested in gaining more prestigious qualifications or acquiring an education in English. Students are a significant portion of the world's mobile population.

This mobility has, however, also produced a response from authorities concerned about the impact of the movement of students into their societies. The worries range from the burden that their numbers impose on social services (from transport to healthcare), to the competition they add to the labour market, to the danger that students will literally

---

[15] Smith, A. (1981). *An Inquiry into the Nature and Causes of the Wealth of Nations.* Indianapolis: Liberty Fund, vol. 1, p. 138 (italics added).

[16] See Bhandari, R., and Blumenthal, P. (2011). 'Global student mobility and the twenty-first century Silk Road', in Bhandari, R., and Blumenthal, P. (eds.), *International Student and Global Mobility in Higher Education: National Trends and New Directions.* New York: Palgrave Macmillan, pp. 1–24, p. 1.

overstay their welcome and turn into long-term or permanent immigrants. If the point is to reduce immigration, students are a problem.[17]

The consequences for students of this view taken by governments have included rises in the cost of studying (in part because of higher fees for visas), increased scrutiny through more complex and time-consuming applications for admission, restrictions on rights to work, and a greater risk of exclusion or deportation for visa violations. But the subject that merits further examination is the consequences for the education sector in the receiving societies.

The most obvious consequence of stricter immigration controls is a rise in costs for schools and universities. Higher education institutions in particular have to devote a greater part of their human resources budgets to employing people whose job it is to ensure compliance with changing immigration laws. This has meant not only recruiting staff to monitor students and faculty to ensure that they have secured and maintained the correct immigration status, and engaging firms specialising in immigration law to make sure that legislative and administrative changes are not missed, but also imposing additional responsibilities on faculty and other professional service staff who might be obliged to report on students. In Britain this has led one senior university official to observe that 'in effect, our universities are now acting in conjunction with the UK's Border Agency to manage immigration'.[18] The financial risk associated with

---

[17] The problem, of course, is how to count students in the immigration statistics. Since students typically are people who intend to come to study and then return home – and usually are obliged to do so under the terms of their visas – they should not count as immigrants. However, given that they often come to study for longer than a single year, the duration of their stays makes them technically 'immigrants'. Whether or not they should be counted in the official statistics is a matter of political dispute. At the time of writing this, in the United Kingdom, senior members of the Conservative government were not in agreement on this question. The problem is further compounded by the unreliability of the statistical methods used for recording and counting students resident in the United Kingdom. See Swinford, S. (2017). 'Immigration figures under review as new checks suggest that numbers are far lower than thought', *The Telegraph*, 23 August 2017, www.telegraph.co.uk/news/2017/08/23/immigration-figures-review-new-checks-suggest-numbers-far-lower (accessed 3 October 2019).

[18] Dean of the Faculty of Arts at Kingston University, quoted in Brooks, R. (2015). *The Impact of UK Immigration Policies on Students and Staff in Further and Higher Education*, Report of the University and College Union (UCU), p. 16, http://classonline.org.uk/docs/Impact_of_UK_immigration_policies_on_students_and_staff_in_further_and_higher_education.pdf (accessed 3 October 2019).

not cooperating is being subjected to fines or loss of sponsorship licences (which would make it impossible for the college or university to admit international students[19]).

No less significant a concern is the impact of immigration control on faculty recruitment. Universities and colleges, and increasingly schools, operate in a global market. Immigration controls impose not only financial costs but also burdens that are the result of delays or uncertainty – making planning more difficult and service provision more unreliable. Thus, for example, in the United Kingdom in 2016 there was a reported shortage of high school teachers in science, mathematics, computing, and selected languages that remained unaddressed in part because immigration regulations did not specify those specialisms on the Shortage Occupation List. Immigration law further required that foreign employees had to meet the minimum salary threshold of £35,000, making it more difficult still to recruit.[20] The Shortage Occupation List is regularly revised on the recommendations made by the independent Migration Advisory Committee, but this information is not available until shortages have already emerged.

The same sort of problem arises wherever immigration restrictions prevent timely recruitment. For example, in the United States in the 1990s a rapid rise in demand for computer programmers went unmet because of the limited number of H1B visas available. The United States government responded by increasing the number of visas, only for them to become available just at the time of the so-called 'dot.com crash' that led to a downturn of demand for programmers and a large number of unused H1B visas. The government responded by reducing the number of visas available, just in time to see demand increase again when the market recovered. The impact here was felt less by would-be immigrant programmers, whose skills were in demand everywhere, than domestically by

---

[19] In 2011 in the United Kingdom 450 colleges, or one-fifth of the further and higher education sector, had their sponsorship licences revoked.

[20] Pells, R. (2016). 'Science teacher shortage spreads, forcing government to relax immigration restrictions', *The Independent*, 26 January 2016, www.independent .co.uk/news/education/education-news/science-teacher-shortage-government-relax-immigration-restrictions-foreign-education-department-a7547406.html (accessed 3 October 2019). The article also reports that the government had missed its own teacher recruitment targets for the previous four years.

individuals and industries whose freedom to recruit was curtailed by immigration controls.[21]

While costs of these sorts are significant, however, it is perhaps more troubling still that a necessary concomitant of immigration control is the increase in monitoring and surveillance of educational institutions by agencies of government, bringing about a decline in their autonomy as well as a transformation of their operations. It means governments requiring education professionals to conform to standards or comply with directives that have no bearing on their own missions or judgements about how their purposes are best served. It also means turning these institutions into arms of government by requiring them to monitor and report on people for whom they have responsibility, establishing within schools, colleges, and universities a norm of continual mutual surveillance. At the extreme, it threatens to bring about an active presence of the police as law enforcement agents – which is currently a live issue in American schools.[22] It may be for this reason that we have seen schools (as well as cities) declare themselves to be 'sanctuaries' and decline to cooperate with immigration enforcement agencies[23] – citing, among other things, harms to children and families, particularly in immigrant communities, traumatised by the fear of police action.[24]

The broader point, however, is that immigration control in the sphere of education has wider implications. Aside from the direct costs it imposes on the institutions themselves, it has an impact on the constituencies that sustain them. This is true not only of the education sector but also of civil society more generally.

Whether it is attempted through the control of employers or the control of institutions in civil society more generally, all efforts to control

---

[21] I am grateful to Dan Griswold for this example.

[22] Keierleber, M. (2017). 'Trump order could give immigration agents a foothold in US schools', *The Guardian*, 22 August 2017, www.theguardian.com/us-news/2017/aug/22/trump-immigration-us-schools-education-undocumented-migrants (accessed 3 October 2019).

[23] Keierleber, M. (2017). '"Sanctuary schools" across America defy Trump's immigration crackdown', *The Guardian*, 21 August 2017, www.theguardian.com/us-news/2017/aug/21/american-schools-defy-trump-immigration-crackdown (accessed 3 October 2019).

[24] One reason for this fear is the frequency of enforcement error. Another is that children with one 'undocumented' parent run the risk of being separated from families.

immigration must affect the personal lives of citizens and residents in that society, though some will be affected more profoundly that others. All must share in the economic cost, but a number will face its implications more directly. To understand this point more fully, it would be useful to consider a couple of concrete cases. For example, Paul and Gail Freahy, a native-born British man and his South African wife, sold everything and moved to Britain from South Africa after having been the victims of violent crime, to open a small photography shop in Torquay. Despite having been married for 13 years, they were refused a renewal of a spousal visa because, as shop-owners, they had not paid themselves salaries and therefore failed to meet the eligibility criteria. Gail Freahy, the non-British partner, was then ordered to leave within 28 days. There is no possibility of appeal from within the United Kingdom, and an application for a different visa would have to be made from South Africa. Deprived of the £55,000 annual income from their jointly run business and faced with the cost of returning to South Africa as well as finding the funds to apply for a visa, the couple were financially ruined since separation was not economically viable. The burden of immigration control here has fallen as heavily on a native Briton as it has on his would-be-immigrant wife.[25]

Or consider the case of Johann 'Ace' Francis, who was born in Jamaica but grew up mainly in the United States. Because his step-father was in the military, the family moved regularly but settled in Washington state when he was seven years old. His mother became a naturalised American citizen when Ace was 14, making him automatically an American citizen also. His biological father had never had custody of him and was not listed on his birth certificate. Shortly before he turned 18, Ace's mother moved to Georgia, but he remained in Washington state to finish high school. During spring break before graduating he visited the town of Seaside in Oregon, where he was caught up in a street fight. He was arrested and, having said he was born in Jamaica and being unable to prove his American citizenship, was deported after being moved first to

---

[25] Shaw, N. (2017). 'Couple given 28 days to leave UK after visa denied by Home Office because they own a shop', *Daily Mirror*, 1 February 2017, www.mirror.co.uk/news/uk-news/couple-given-28-days-leave-9734476 (accessed 3 October 2019).

Las Vegas and then to a detention centre in Arizona. After being deposited at Kingston airport, Ace lived on the beach and streets until he found his biological father on Jamaica some time later. His mother had no way of knowing that her American son had been wrongfully deported. His father contacted his mother but it was 10 years before he was eventually returned to the United States. His mother wrongly assumed he must have been deported because he was not naturalised. When they figured out that he had been naturalised automatically, it took several years to find the necessary records, including birth certificates, to obtain the papers needed to get an American passport and return home. Ace Francis died shortly after his return.[26]

Such stories can be multiplied 1,000-fold – indeed a million times over, given that more than a million American citizens have been wrongly deported since 1930. In the United Kingdom the government estimated that its visa rules would *each year* prevent about 18,000 British citizens from being united or re-united with their spouses or families because of the income threshold that had to be met in order to sponsor a partner to enter the country – to say nothing of those denied that freedom for other reasons.[27] Since more than 40% of those in employment, and 55% of women in the workforce, do not meet the threshold there is a sad irony in these figures to the extent that one of the express purposes of immigration control is to protect the poor from the economic ill-effects of the influx of outsiders.

The Apartheid Analogy

Immigration control is the result of an attempt to resolve a dilemma. The authorities of states around the world have always confronted a problem

---

[26] The full story is recounted in the Preface to Lawrance, B. N., and Stevens, J. (eds.) (2017). *Citizenship in Question: Evidentiary Birthright and Statelessness.* Durham: Duke University Press, pp. ix–xiii.

[27] A citizen or permanent resident needs to prove having an income of at least £18,600 to sponsor a husband, wife, or civil partner, or an income of £22,400 for families with one child (and £2,400 for each additional child). The income of the sponsored spouse cannot be taken into consideration. The £18,600 threshold is the point below which a British citizen or resident becomes eligible for welfare benefits.

posed by a combination of social transformation and the need to adjust to new circumstances, on the one hand, and a mobile population both within countries and abroad, on the other. It seems that at times there are too many people, especially workers, and at other times just not enough. But controlling matters through managing immigration is not easy because, in the end, people are not like pieces on a chessboard that can be moved about at will. Moreover, they usually come not as abstract units of labour, or as potential citizens of a fixed type, but as human beings with their own baggage, including their families, their duties to those left behind, and their tendency to form attachments and relationships wherever they settle. Controlling such people requires a serious commitment which, if taken to its furthest extremes, can result in a serious loss of freedom for immigrants and citizens alike.

Here something might be learned by considering one of the most interesting cases of social control in the twentieth century: the example of Apartheid South Africa. For a variety of reasons, white South African governments from the beginning of that century did not want black Africans to be a part of white society, even as they felt that they needed to draw upon black labour to further the goal of economic development. Taxation of black enterprises and policies of land reform that made black farms less viable led to greater availability of cheap black labour. However, the influx of poorly paid black Africans into the cities put pressure on white municipalities that needed to supply housing and social services, with the result that policies of segregation were tightened – forcing more and more workers to live in segregated townships or to return to their distant homes and families each day. In effect, black Africans were 'forced into a system of migrant labour'.[28] From the point of view of the white government, the problem was how to maintain the supply of cheap labour without any loss of convenience, and also without the establishment of a firm black foothold – economically, culturally, or politically – in European society. The policy and ideology of Apartheid were designed to solve that problem. Apartheid was South Africa's internal immigration policy.

---

[28] Clark, N. L., and Worger, W. H. (2004). *South Africa: The Rise and Fall of Apartheid.* London: Longman, p. 22.

The enforcement of this immigration policy required laws. Besides legislation to control the employment of workers, the South African government also introduced measure to control other aspects of life that might threaten segregation: the Prohibition of Mixed Marriages Act (1949), and the Immorality Act (1949) outlawing sexual relations across racial groups, were early examples. These Acts, however, meant that there was now a need to pass the Population Registration Act (1950), which classified South Africans into four racial types – black, white, coloured, and Indian (the latter two groups being further divided into sub-classifications). But maintaining control required more than the passing of laws. It also necessitated the forcible movement of populations, and thus the removal from their homes of 3.5 million non-white South Africans between 1960 and 1983 – many of them 'deported' to designated tribal homelands or Bantustans and declared to be not (or no longer) South African citizens. It necessitated the introduction of 'Passbooks', which black Africans moving through the country, but particularly through cities, needed to produce for inspection on demand by white authorities, in particular by the police.

The aspiration of the National Party at this time was a conception of white rule that saw total segregation as the ultimate objective, with a slow move towards the elimination of Africans from industry until blacks came to be no more than visitors to cities to meet occasional labour needs. The labour of Africans would be managed by government bureaucracies, with the black population able to move about the country only under government supervision.

Turning this vision into reality, however, was another matter altogether. First of all, it required the compliance of the black African population, which had little incentive to embrace either the laws or a way of thinking that ran so clearly against their interests. Thus, a good part of the policy of apartheid, as Gwendolen Carter argued, was about trying to 'make the African the different kind of person that theory says he is',[29] – the 'theory' being the theory of Apartheid. It was this aspect of Apartheid against which the South African activist Steve Biko struggled, arguing

[29] Carter, G. M. (1958). *The Politics of Inequality: South Africa since 1948*. New York: Praeger, p. 15.

that it sought a transformation of consciousness, to make black Africans see themselves only as the system viewed them and never as free and independent human beings. But a transformation of the black African consciousness was never going to be enough if the ideology of Apartheid was not secure within the white South African population, so an equally important condition of its preservation was the control of society more broadly. That meant control of any recalcitrant elements of the white population.

The brutality of Apartheid as a system of social control is most readily visible in its treatment of the black population. Aside from the systematic injustice they suffered under laws that limited their rights and freedoms, black South Africans were the victims of repeated acts of physical violence perpetrated by the police and the military arms of government. But Apartheid also required the gradual escalation of control of – and violence against – the white population whenever it showed any sign of wavering in its support for government policy, and for the policy of Apartheid in particular. Under the Native Laws Amendment Act (No. 54) and the Abolition of Passes and Coordination of Documents Act (No. 67) in 1952, regional passbooks were abolished and replaced with 'reference books' containing vital information including taxation and employment records, and Africans travelling without their reference books were subject to criminal punishment, including imprisonment. In 1970 similar books were issued to other races, including whites, under the Population Registration Amendment Act. White employers were already forbidden to hire people who had not been appropriately vetted by the government. Increasingly, however, the authorities clamped down on any form of dissent that appeared in the press, in popular culture (that was highly censored), and in the courts. By common consent, South Africa developed into a police state, but one that exercised its power not only over its black majority population but also over the minority of whites that might have threatened the stability of the regime.[30] It depended for its survival on

---

[30] For a famous account of the lengths to which the South African government went, see Woods, D. (1987). *Asking for Trouble: The Autobiography of a Banned Journalist*. London: Penguin (first published 1980). See also the film by Richard Attenborough, *Cry Freedom*, dealing with the death of Steve Biko and the flight of Donald Woods from South Africa.

the acquiescence of all those subject to its laws, and that required enough people buying into the ideology that sustained it. Members of the white population who dissented in public thus had to be silenced. Indeed, such people were not only forbidden to speak out in South Africa, but also prevented from travelling abroad in case they spoke out against the government and criticised the policy or the ideology of Apartheid.

Immigration Control and Equality

Controlling immigration requires policies and institutions of control that control citizens and residents no less than immigrants or would-be immigrants. To be at all effective, however, it requires control not only of movement but also of a large range of activities in social life and, ultimately, a measure of control over how people think. People are most readily controlled when they believe the controls that constrain them are warranted, or when they no longer recognise such controls as limitations at all. Yet immigration controls reach even deeper into the values that distinguish liberal democratic societies, for they also have a profound impact on equality and the rule of law.

The most obvious reason for the troubling impact of immigration control on equality is that controlling immigration invariably requires *selecting* people for admission or inclusion and so treating them unequally. The history of immigration control makes this very clear, since everywhere the attempt to control movement has been an effort to allow some people in and keep other people out. Immigration control is a danger to equality because its pursuit threatens to undermine the egalitarian ethos of a free society, as well as to distort its institutions and ultimately make the sustaining of equality before the law attainable only inadequately, if at all. There are several reasons for this. Most obviously, immigration controls emphasise the importance of the differences among people, with a view to justifying treating some more favourably than others. First, it means distinguishing insiders or natives from outsiders or foreigners – with a view to privileging the former over the latter. Second, it means distinguishing among foreigners, for immigration control invariably turns into selective immigration that prefers some kinds of people to others. Third, it means distinguishing among the citizens or residents of one's own society, and privileging some among their number over others. No less importantly, immigration controls encourage the development of institutions, both within government and in civil society more generally, that distinguish among people on the basis of their nationality and the traits that make some persons likely to be granted the freedom to immigrate. Immigration controls encourage or give succour to those elements in society that look to protect the privileged standing of some sections of society.

To understand the inegalitarian spirit that underpins the imperative to control immigration, it is easiest to begin by returning to its history. Any turn to the history of immigration control brings us rapidly to the matter of race and racism. Immigration control in the liberal democratic West has, to a significant extent, been about limiting the entry of, or keeping out altogether, people of the wrong ethnicity, religion, colour, or (more recently) culture. The most obvious example of this might be thought to be Australia, whose 'White Australia Policy' advertised its intentions with striking candour. Yet it was hardly alone in this regard. In some ways, the history of America's immigration policy is equally, if not more, instructive, since it reveals how debates over immigration shaped the country's racially inflected thinking about equality, the role of political institutions, and what it could mean to be an American. Similarly, the history of British policy, particularly in the twentieth century and in the years since decolonisation and the loss of empire, reveals a tightening of immigration controls – controls that fall more heavily upon peoples from Asian and African countries. Even the European Union, despite its enthusiasm for freedom of movement within its own external borders, and indeed its formal commitment to human rights norms and conventions that proclaim the importance of racial equality, has hardened its resolve to resist the encroachment upon its territory of peoples to the south and east of its geographical boundaries.

But can there not be a non-racially inflected immigration policy? If we leave aside the extremes of immigration policies that are either completely restricted (permitting *no* entry into the state) or completely *un*restricted, it is very difficult for governments to leave race out of the equation. This is not to say that every immigration policy must be racially motivated or biased to the same degree. But for as long as immigration is *selective* rather than simply randomised it is difficult for policy not to be racially biased to some degree since some kind of selection criteria will be applied.[31] If the point is to distinguish, first, between insiders and outsiders and, second, among different outsiders, some reasons will need to be offered for choosing certain people and rejecting others. Race or ethnicity will turn out to be a salient marker for exclusion.

More than this, immigration control is one of the most important drivers of racial differentiation. The most important reason to think that immigration control is a likely contributor to racial differentiation is that racial differences have no natural basis but are the product of needs or incentives to distinguish and classify persons for any of a variety of purposes. Controlling immigration has never been simply about limiting the *numbers* of persons entering a society but has always been about the *kinds* of person who are admitted, so the practice of immigration control is everywhere a practice of selection.[32] Once selection is the name of the game, differentiation, classification, and comparison follow. No less importantly, once distinctions are made they have to be sustained, either by explicit defence of the categories created, or by more subtle means like using proxies for distinctions that cannot be drawn in the open. Immigration control – through immigration law and immigration enforcement – has contributed substantially to the creation of racial categories. It would be going too far to suggest that it is the only source of racial differentiation, but its significance should not be underestimated.[33]

---

[31] The 'Green Card Lottery' used by the United States might be one example of a randomising system of admissions. However, it is worth noting that the first 'diversity visas' issued in 1987 were awarded on a first-come-first-served basis, favouring those who first knew about the opportunity – primarily those from Canada, Ireland, and the United Kingdom, and with the Irish getting 40% of the visas. Between 1991 and 1994 the programme was amended to favour Europeans. In this period 40% of visas were explicitly reserved for Irish applicants.

[32] Whenever purely numerical limits of immigration are set by governments, different groups invariably complain and petition for exemptions – employers because they want workers, communities because they have ties abroad, individuals because they want family reunification.

[33] There is a substantial body of literature on the emergence of racial differentiation as a phenomenon with roots in particular interests and political imperatives in the nineteenth and twentieth centuries. See, for example, Lake, M., and Reynolds,

The broader conclusion to which all this points is the intimate relationship between immigration control and racialisation. It is well established that immigration law, as well as enforcement practices, are significant forces in the construction of race – in the United States and elsewhere.[34] It should also be clear that this not only affects immigrants but also 'generates a host of *practices* that redound to the disadvantage of those citizens who share characteristics with immigrant communities'.[35] To the extent that any society aspires to develop on the basis of racial equality, immigration control is a part of the problem rather than a solution.

No less troubling from the point of view of equality is that it also encourages the development of policies or institutions that compromise the rule of law. Once again, history supplies ample evidence of this tendency. For example, numerous legislative enactments would seem to be contrary to the spirit of the rule of law. Though cast as general laws, they are in many cases crafted so as to apply to specific classes of people, and, while they may once have been internally consistent, they were much less consistent over time, and too often were amended to suit short-term concerns about particular policy outcomes. Those subject to the law were not considered equally but discriminated against on the basis of health, disability, nationality, and race – sometimes explicitly, but often implicitly, though by design. Institutionally, the rule of law was compromised to the extent that the right of appeal and access to courts was limited. In practice, it was further disregarded insofar as the law was

H. (2011). *Drawing the Colour Line: White Men's Countries and the International Challenge of Racial Equality.* Cambridge: Cambridge University Press; and Allen, T. W. (2012). *The Invention of the White Race. Volume I: Racial Oppression and Social Control* and *Volume II: The Origin of Racial Oppression in Anglo-America,* 2nd edn. London and New York: Verso.

[34] For a survey of some of the relevant literature, see Chacón, J. M., and Bibler Coutin, S. (2018). 'Racialization through enforcement', in Bosworth, M., Parmar, Alpa, and Vázquez, Y. (eds.), *Race, Criminal Justice, and Migration Control: Enforcing the Boundaries of Belonging.* Oxford: Oxford University Press, pp. 159–75. For the United Kingdom, see, for example, Miles, R. (1990). 'The racialization of British politics'. *Political Studies,* 38(2), 277–85; and Joshi, S., and Carter, B. (1984). 'The role of Labour in the creation of racist Britain'. *Race and Class,* 25(3), 53–70.

[35] Chacón and Bibler Coutin (2018), 'Racialization through enforcement', 161.

not applied equally but used to target particular groups while sparing others.

There are three ways in which the rule of law is compromised or eroded under pressure to further control immigration. First, changes are made to the content of the law to make it better suited to serve the ends of policy, whether or not that is consistent with the demands of legality in any higher sense. Second, the institutional arrangements that make law accessible and officials subject to scrutiny are (further) revised to reduce their capacity to slow or check the implementation of policy. Third, the pursuit of the objectives of policy is simply conducted in violation of the law, since, even if some failures of compliance are brought to attention, many will go undetected and the objectives of policy will be closer to being realised.

The deep source of the problem, however, is not the misbehaviour of particular officials, for one has to ask what it is that brings political leaders and public servants in liberal democracies to take steps that are so out of keeping with long-established traditions. The answer has to do with the aspiration to acquire and retain a measure of control over immigration that is, in the end, unattainable – at least if such societies are to remain ways of living in which the ideals of freedom and equality hold sway. The pursuit of the ideal of control means that every setback will have to be met with a search for other measures to secure it, and as the law reveals itself to be an obstacle to its achievement so will the law be changed, worked around, reinterpreted, or ignored. If a government sets a target of 100,000 net immigrants a year and finds itself completely incapable of meeting it, the temptation to work around the law may become too great to resist. At every step in the ladder of authority and responsibility, people will be under pressure to meet their own targets; and the more forceful the imperative the greater the pressure and the more likely the yielding to temptation to give short shrift to the spirit of the law.

The Case for Control: Economics and Culture

Even if these arguments raise important concerns, however, it might nonetheless be argued that the risks to freedom and equality are worth taking because there are some important gains to be made by controlling immigration. More precisely, immigration control is necessary because it brings with it substantial economic benefits, and also makes it possible for a society to preserve its cultural integrity. If trade-offs must be made, freedom and equality must give way to economic necessity and cultural integrity. Let us consider these two arguments more closely.

The economic argument has two aspects, the first concerning the aggregate impact of immigration on a country's wealth and the second the distributive consequences of immigration. The first concern, that an influx of immigrants will diminish a country's wealth and overall well-being by forcing natives to share the economic pie with more and more people, is widely acknowledged to be unfounded. While immigrants consume, they also produce: more mouths to feed also means more hands to work. In the short to medium term the economic consequences are more or less neutral. Though it is difficult to find agreement on the finer points of the methodologies used to measure impact, the consensus is that there is no reason to think immigration produces either a substantial drop or a significant increase in national wealth or income in the short term. In the longer term, the gains are undoubtedly positive, though once again the margins are small. All this does not take into account the impact of immigration on innovation and the contribution it makes to the dynamism of an economy.

The more substantial worry, however, is that whatever economic gains might come from immigration, the benefits are enjoyed by the most well off in society while the losses are borne by the poor. An influx of cheap labour lowers prices, but it also lowers wages and creates a pool of unemployed workers – disproportionately among the native population, which is not able to take on jobs at substantially lower wages. Even if in the long run everyone gains, in the short run the price is paid by the most vulnerable parts of society.

Here it ought to be acknowledged that, with the exception of those with particular skills not found (or in very short supply) in the population of the host society, every immigrant who finds work takes a job from a native who might otherwise have been employed. But this is not enough to show that immigration overall reduces native employment or lowers native wages. First, while immigrants take jobs, they also generate them.

If their skills mean that they are in effect *substitutes* for native labour, they compete with native workers. But if their skills are *complementary* to those found in the native population, they help to generate employment by making some enterprises economically feasible or more profitable. Second, by enlarging the size of the domestic market, immigrants also generate demand for goods and services, further stimulating production and employment. The overall consequence of this is positive in the longer term, and neutral in the shorter term.

That said, it would be too much to claim that there could never be any losers from immigration. Even if wages did not fall as dramatically as some fear or employment did not decline precipitously, *some* workers will see their wages fall, and others may lose jobs because of immigration. Though those numbers may be small, from the point of view of those individuals the outcome may be hard – or even harsh. Yet what must also be recognised is that similar consequences can also follow from a loss of workers arising from reduced immigration. Some domestic industries depend for their success on the availability of immigrant labour. This may be because certain skills are not to be found in sufficient supply within a country. Or it may be that the domestic labour supply is not mobile enough, since unemployed workers may not be able to move from one end of the country to the other to take up employment opportunities – particularly if it means separation from family or asking a spouse to give up his or her own job. If firms cannot find workers, they may cut back or close, to the disadvantage of other native workers. In countries like the United States, Australia, and the United Kingdom, a small proportion of the population do indeed suffer a reduction in wage income as a result of immigration, but protecting this group of people through immigration control may not do anything more than transfer the burden of loss to other native workers and employers who depend upon immigration for their own livelihoods.

One other point is worth noting in considering the economics of immigration. In the long term, immigration has a positive effect overall on a nation's wealth and income, so future generations will reap the benefit of a more dynamic and productive economy. If the argument is that immigration control is necessary to protect the interests of native workers who might suffer a loss of employment or a reduction of income, then why should *current native* workers be protected at the expense of *future native* workers? If the purpose of immigration control is to defend the interests of natives, it has to be asked *which* natives. Immigration control may harm some of them, but it also benefits others. It does not look like there is any obvious way of determining which natives are more deserving.

The economic arguments for immigration control are difficult to sustain. But the cultural argument is of a different order, and perhaps this is where the case for control rests on firmer ground. The argument in a nutshell is this: an influx of any significant number of immigrants, particularly from countries with very different traditions, threatens to undermine the cultural integrity of a society. A society, after all, is not just an economy but a political community, with its own traditions and way of life, and while it can tolerate a limited amount of movement of people into its territory, this has to be managed carefully to ensure not just the prosperity but the very survival of a people. There is surely something to this.

While it seems evident that the sudden infusion of a massive number of new settlers might be troubling or even harmful to a society that is not ready to cope with dramatic changes, it should be noted at the outset that the issue here is not simply one of numbers. The cultural argument is not concerned so much with *how many* immigrants arrive as with *who* those immigrants might be. Invoking the cultural argument may make sense for culturally homogeneous societies, particularly if they are small, but it poses a dilemma for liberal democracies that are already marked by a measure of cultural diversity. These include most especially the societies of Britain and western Europe, which are internally diverse as a consequence of their colonial traditions and histories of political expansion and conquest, North America with its indigenous and slave populations as well as its own more recent histories of warfare and the incorporation of peoples, and Australasia, with its indigenous peoples and attempts from time to time to increase its population through immigration when labour was scarce. The dilemma they face is the result of the fact that they are societies that are already diverse, combined with their political commitments to freedom and equality.

In a diverse society, cultural integrity is difficult to preserve because the very definition of the national culture is difficult if not impossible to settle. The reason why it needs to be settled is that immigration control can only help preserve cultural integrity if it is selective. Admitting people on a randomised basis leaves the matter to chance, so deciding who to include and who to discourage is vital. But this creates a number of problems for a diverse society committed to freedom and equality. First, it requires a judgement about whom to include and whom to exclude *within the existing society.* If the nation is to be viewed as possessing a particular cultural character, then which culture is the national culture – and which cultures are to be consigned to the periphery? Second, however this question is answered, it means dividing society into different categories of people who will not then share equal standing with one another. Moreover, the admission criteria for future immigrants will then discriminate against the wishes and interests of those on the cultural periphery as they are advised that people of their type are less welcome to immigrate or are perhaps to be excluded altogether.

If immigration is to be controlled in the interests of nationality, and nationality is culturally defined, then the boundaries of nationality will have to contract. This is precisely what happened in the British case when the country divested itself of its empire and over a generation took away the nationalities of millions of its former subjects. It was a dilemma well recognised at the time, if only by some. The Conservative MP Sir David Maxwell Fyfe expressed this concern with respect to the creation of a different forms of citizenship within the British Empire in the discussion in Parliament during the Second Reading of the British Nationality Bill of 1948. He said that

> We deprecate any tendency to differentiate between different types of British subjects in the United Kingdom. We feel that when they come to the United Kingdom there ought to be an open door and a reception for every type. If we create a distinctive citizenship for Britain and the Colonies, inevitably such differentiation will creep in. We must maintain our great metropolitan tradition of hospitality to everyone from every part of our Empire.[36]

Fyfe's argument was that any effort to distinguish between British and Colonial citizens within the Empire could only mean a kind of

---

[36] *Hansard,* 453 (1948), col. 411.

47

differentiation among British subjects that was inconsistent with the principles that had been established by the Imperial Conference of 1930 – that all British subjects would hold British nationality, and that that nationality could not be taken away by the actions of any power. Differentiating citizenships, he feared, would not only reduce various protections and rights British nationals throughout the colonies enjoyed but also mean treating them unequally.[37]

The alternative to construing culture narrowly and excluding portions of the population from the recognised cultural identity would be to define culture and national identity more capaciously. Here the idea is not to push some citizens to the periphery but to develop a more inclusive conception of cultural identity. On this account, a culture is not such a homogeneous construction but more like a river fed by many tributaries. But while there is much to be said for this way of viewing the world or a national society, it does not help the argument for immigration control from the perspective of cultural protection. If a culture is broad and inclusive, what would be the basis of selective immigration control – and, indeed, what would be the point?

Taking Back Control

While these various arguments drawing attention to the problems of immigration control may have a point, however, there is one final consideration that deserves attention. This is the idea that immigration control is necessary because, ultimately, it is important that a nation take control of its own destiny. Homogeneous or diverse, it is in the end one nation, and the most significant ability it has to determine its own future is the ability to control its borders and decide who is to be a part of that society. Rhetorically, this has been the most forceful argument advanced in support of immigration control: the language of a national community

---

[37] He notes in particular that with respect to the 'racially distinct and smaller countries of the Commonwealth' Britain should take pride in having 'imposed no colour bar restrictions making it difficult for them to come here' and ensuring that those who came to Britain 'found themselves as privileged in the United Kingdom as the local citizens. See *Hansard*, 453 (1948), col. 403.

'taking back control' is the language that resonates most powerfully in public debate.

Yet while the rhetoric is arresting, the arguments themselves are less convincing, for they face a number of difficulties. Most obviously, it should be noted that, despite the longing for national control, there are limits in the modern world to the capacity of most nations to go it alone. Controlling immigration is not a matter of exercising power unilaterally. It is everywhere a collaborative endeavour among nations that have created institutions to govern population movements, and the signatories to the regimes and international agreements are bound in various ways. A nation cannot deport an individual without the agreement of the country to which he or she is to be sent, and a refugee cannot legally be returned to a place of danger – at least not without violating the spirit of the law.

Even to the extent that nations do have a capacity to determine who comes and who goes, however, it has to be acknowledged that many liberal democratic societies are divided on the question of immigration – even if not evenly divided. This means that any decision to move in one direction or another means moving in a direction that some people do not want. To be sure, this may be unavoidable, and in a liberal democracy most people accept that they may not get their way. But it would be too much to assert that this means self-determination in anything more than the most abstract sense of the term.

It further bears recognising that if self-determination is what really matters, it is also quite possible to take back control by reducing immigration controls and taking a more liberal and open attitude to the movement of people and the inclusion of others. This is an option that has not been sufficiently considered by the advocates of self-determination, but is worth another look. If there is any truth in the argument advanced here that immigration control ultimately involves not so much control over others – foreigners or outsiders – but control over ourselves then an act of self-determination to reduce this burden should surely be welcomed by anyone who cares about freedom. Indeed, if immigration control means imposing greater controls upon ourselves, why would we wish to call it self-determination?

3 Art and Migration

On the Power of Movement, Light and Shadow

KHADIJA VON ZINNENBURG CARROLL

This chapter, which is divided into two parts, includes the script of the performance lecture *Men in Waiting*, preceded by the research on art and migration that underpins it. The project is in part a phenomenology of incarceration that studies what happens to perception when it is limited to the prison architecture used for immigration detention in the United Kingdom. It is also an artistic enactment of that subjectivity, and this text is a reflection on the time and space produced in this performative reflection. Through puppetry the performance takes place in a shadow world, not unlike that of Plato's protagonist in the *Republic* who seeks to bring the people to enlightenment and is made a martyr as a result. Plato's allegory would indicate that we all have access to the human condition of confinement in the dark with only shadows. However, the specific shadowside of the world as seen through the United Kingdom's immigration detention centres was what this Darwin Lecture immersed the audience in. Writing and performing this play was a way to sit with the shadow, and within the shadow.

Living in Darwinian Shadows

The extreme sensitiveness of certain seedlings to light, as shown in our ninth chapter, is highly remarkable. The cotyledons of Phalaris became curved towards a distant lamp, which emitted so little light, that a pencil held vertically close to the plants, did not cast any shadow which the eye could perceive on a white card. These cotyledons, therefore, were affected by a difference in the amount of light on their two sides, which the eye could not distinguish. The degree of their curvature within a given time towards a lateral light did not correspond at all strictly with the amount of light which they received; the light not being at any time in excess.

50

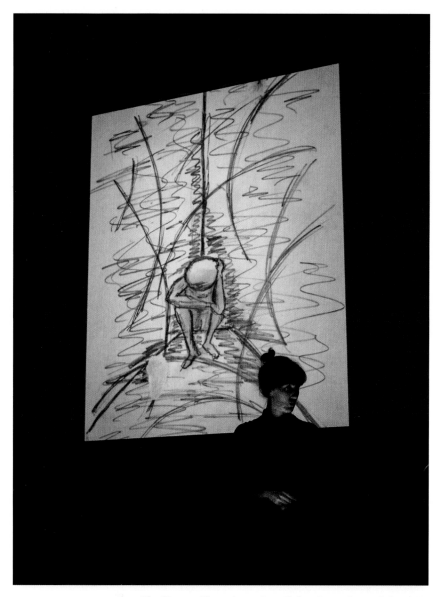

FIGURE 3.1 Khadija von Zinnenburg Carroll showing a charcoal sketch by an unnamed detainee in Immigration Removal Centre Campsfield House about how it feels to be in detention, from the Immigration Detention Archive Oxford drawings in Darwin College Lecture by Khadija von Zinnenburg Carroll, Cambridge, February 2018. Photograph by Sircam.

They continued for nearly half an hour to bend towards a lateral light, after it had been extinguished. They bend with remarkable precision towards it, and this depends on the illumination of one whole side, or on the obscuration of the whole opposite side. The difference in the amount of light which plants at any time receive in comparison with what they have shortly before received, seems in all cases to be the chief exciting cause of those movements which are influenced by light. Thus seedlings brought out of darkness bend towards a dim lateral light, sooner than others which had previously been exposed to daylight.

Charles Darwin, *The Power of Movement in Plants* (1880), p. 260.

Towards the end of his life, Charles Darwin closely observed the ways in which plants grow to the light. He measured the curve, the precision of their ability to sense light, which the human eye could not even perceive as a shadow. Darwin was studying power in the movement of plants rather than power in the movement of people, but his observations might read, anachronistically at least, as a metaphor for migration. He describes the ways in which the body that is so sensitive reaches towards a light source that we cannot even perceive, that has no shadow in our eyes, but has enough power to draw life into a curving arc.

In that act of bending to the light that Darwin wrote about, the body deforms itself in order to offer a surface that can be penetrated, synthesised, that will photosynthesise and sustain life. In that movement, which often becomes grotesque in its bended contortions, the body strives for growth despite an engulfing shadow from which it must constantly escape. To some the very effort of these bends has a beauty, a skill. In the effort to adapt to the world of shadows in which only a little bit of light enters, a contortion occurs in bodies that have to adapt to compromised light conditions. The body re-members the bends, as do the offspring of these bodies. They will develop defences to survive this world as they reach away from the shadows. The surfaces they hold out are vast and growing, full of details, with the pattern of spines still impressed on their surface. With the marks and traces of their stretch, not all limbs reach as fast as others and only collectively do they support this body that projects itself outwards. That is the power of movements that Darwin described. Twisting and bending, falling, rotting, dying, proliferating rhyzomatically, there is nothing more human than this power of movement.

Platonic Shadows

In this art-research the argument is that migrants become shadows and shadows become a portrait of the state institutions that have been constructed to deal with the migration crisis.

The condition of being a shadow is one that Plato famously described in the allegory of the cave. It has since been used as an attack on the arts as illusory, but I will return to the philosophy of shadows to cast light instead from another angle on the situation of art and immigration detention. Since the condition of 'living in shadows' is one I often heard as a description from migrants, it became one trope I picked up in my response as an artist. In my research I looked to transnational philosophies of the shadow world. The story that Plato tells of prisoners trapped in the darkness, observing only a shadow world, is typically taken as an allegorical critique of visual representation. However, the situation he describes can be turned into an analogy of those migrants detained indefinitely in detention. The press that mediates and thereby defends a kingdom against migration is as illusory as Plato's shadows.

Through fieldwork, the European Research Council-funded project 'Subjectivity and Penal Power' investigated the effect of immigration detention on those whose lives are bordered by living in the 'removal centres'.[1] The condition of living through shadows, of seeing the self as a reflection cast on a wall, from a light source that is behind one, is thus an analogy (in my work), for the popular perception of migration. The migrant is figured by the media as a frightening mass of silhouettes without real form, while the actual figures are kept from view in detention and deportation facilities. For studies show again and again that the more contact Europeans and Americans have with migrant communities, the less they see a danger in their presence. It is the shadow presence that produces fear.

---

[1] See von Zinnenburg Carroll, K. (2015). 'Being in detention: media arts at Colnbrook IRC', www.law.ox.ac.uk/research-subject-groups/centre-criminology/centreborder-criminologies/blog/2015/11/being-detention (accessed 10 January 2019).

Shadow Puppets

The position of art in the migration crisis is highly compromised. Using shadow puppets as the medium to express the work I was doing in and about detention centres referenced the restraints within which we were working. In their simple abstraction they were also particularly well suited to the material.[2] There were several reasons for this that I will explore in my reflections on censorship below. The shadow is the image par excellence. The Indonesian puppeteers at the Papermoon Puppet Theatre would breathe life into the puppets (as I do at the beginning of Scene 1). The enlivening of non-human matter through artistic and ritual practice has long interested me in my research into understanding the meaning of objects to people (when the material stands in for other lost relations).[3] Plato's story tells us that we are trapped in the darkness observing only illusions if we remain with the shadows. There is no contour, no touch, and yet the shadows are more us than we are ourselves. That is one of the teachings of the Indonesian *wayang kulit* shadow philosophy.

For the play *Imigrazie* (Bahasa Indonesian for 'immigration') I turned the main figures I had worked with into puppets: the criminologists, the immigration removal centres, the multitude of detained migrants. I made the multitude of detainees (whose stories mirror and amplify each other's the more we hear) into one puppet that had many faces. I turned the recurrent dreams and nightmares into another puppet, a fabulated character that came to be known as 'the hairy angel'. There was a playfulness to this political parody, a truck driver with a snarling grin drove his white van in and out of the detention centre. Even the building was a puppet, whose mouth opened and closed to swallow people whole. The big multitude puppet was terribly unwieldy, hard for me as amateur

---

[2] See Gross, K. (2011). *Puppet: An Essay on Uncanny Life*. Chicago: Chicago University Press. There are a lot of artists who work with shadows. Of notable influence on me were Joan Jonas and Naoko Tanaka.

[3] See, for example, another case in von Zinnenburg Carroll, K. (2017). 'The inbetweenness of the vitrine: three *parerga* of a feather headdress', in Basu, P. (ed.), *The Inbetweenness of Things: Materializing Mediation and Movement between Worlds*. London: Bloomsbury, pp. 25–45.

dalang (puppetmaster) to operate. In the residency at the Papermoon Puppet Theatre in Yogyakarta I learned how precise Asian puppetry traditions are, why the puppets are the scale and build that they are, so the *dalang* is able to operate many at one time, alone.

Since the shadowed locations of immigration detention are obscured and indeed largely unrepresentable because of legal censorship, art offers a unique form of access. Spatial and temporal boundaries are complicated by and collapsed within the artistic process of representation. Having repurposed the shadow to also represent internal states by showing the silhouette of the migrant, I sought to produce affective and embodied evidence.[4] The guiding idea was that the shadow-site produced in the play might be a more vivid representation of subjectivity than documentary photography or bureaucratic evidence of the detention centres. And yet, in all these various types of evidence, collected by social scientists and interpreted by artists, there was interplay and interpretation of the rules and structures of the institutions in which the shadows took shape. For example, the aesthetic of censorship often literally blacks out powerful information, a process in which ironically that redaction also highlights the importance of the censored information. Redaction is animated live in the performance.

The Immigration Detention Archive

We turned to montage in this project, the layers of which echo the doubling multiplicity of the types of images in the archive – hundreds of women, birds, mosques, homes, flags, and weapons – simultaneously covering the absence and the haunting memory of these presences. The collage has historically been used in political work, for instance by the artists Hannah Höch and John Heartfield. The collages produced as part of the Immigration Detention Archive seek to evoke the tragedy of the disjuncture between destination and the present space of dwelling in detention. It is in the acts of association between the disjoined present

---

[4] For the burgeoning discipline of artistic research, see the Society for Artistic Research's journal *JAR* or *OAR* and, in particular, Bosworth, M., and von Zinnenburg Carroll, K. (2016). 'Art and criminology of the border: the making of the immigration detention archive', www.oarplatform.com/art-criminology-border-making-immigration-detention-archive/ (accessed 10 January 2019).

and the drawn fantasy, and the virtual impossibility of bringing the two together in the reality of the migrant, that the tragedy in this crisis becomes evident.

The photograph in Figure 3.2 is of the storage room of the detention centre behind Heathrow airport, the place you deposit your valuables before you are locked up. In one view, these messily stacked items evoke far bleaker collections of belongings piled up in concentration camps. From another, more recent, perspective, the materiality of the bags, made of recycled strips of other plastic bags, has become synonymous with the refugee crisis. The cheap carryalls have a ubiquitous pattern, volume, and resilience. The photograph we have selected reproduces this storage room within a photographic montage in which it is overlain by a drawing of two women. That printed photograph is, in turn, portrayed in another, quite different space, hanging on the wall of the Bonavero Institute of Human Rights in Oxford during an exhibition we made there in April 2018. As this image shows, during the associated conference participants intuitively put their luggage in front of the photograph of migrant belongings.

The portraits of women drawn by men in Campsfield House emerge from the shadow world overly eroticised. Who are these women? And what were they to those who drew them? Some come in the guise of the idealised form, traced from magazines or tattoo pattern books, others from the fantasies of the men who have no access to their loved ones. In these pages, women are placed over photographs of the men's living spaces, and over self-harm reports to evoke the desperation caused by family separation that is also palpable in the drawings. Absent presences haunt those spaces. The objects of desire become exaggerated in the face of the deadening delay of gratification.

In one letter a woman held in Yarl's Wood Immigration Removal Centre describes the London that she imagines as a place filled with flowers and beauty. The unsure greyleaded shape of an English female figure with a hat and gloves emerges scratchily on another page. The unreality of these images is not important, just as the recognisable mimesis of the portraits is not relevant. Every imprint of another from the space of isolation is a way of reaching beyond. In the shadows women appear to become sex objects perhaps because desire has no contour, is overturned, overwrought (there are many such drawings in the archive).

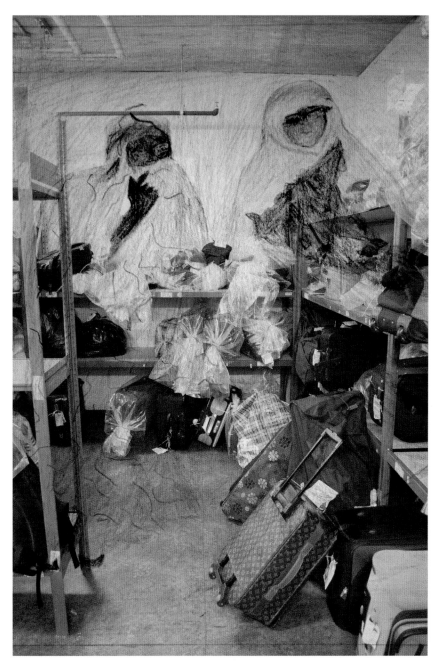

FIGURE 3.2 Unknown detainee (drawing), Khadija von Zinnenburg
Carroll (photograph), Christoph Balzar (collage), photograph of detainee property room
at Immigration Removal Centre Colnbrook overlaid with drawing of two muslimas, 2018.
Courtesy of the artists and the Immigration Detention Archive Oxford.

FIGURE 3.3 Unknown detainee (drawing), Khadija von Zinnenburg Carroll (photograph), Christoph Balzar (collage), car park at Colnbrook and Harmondsworth with floral motif, 2018. Courtesy of the artists and the Immigration Detention Archive Oxford.

The Flowers of Detention

There is a remarkable quantity of flowers in the archive, as the pin board of the art room shows. Some of these drawings have geometrical features from Islamic motifs, for example the decorative themes of flower and bird *gol-o bolbol* and *gol-o morgh* in Persian art.[5] In this kind of Persian art and literature, the flower is a metaphor for the beloved and the bird is considered as the lover or poet. Other drawings were made following flower arrangement classes. The unlikely nature of this activity for a community of men facing deportation raises questions. Is this a deliberate strategy designed to undermine masculinity and strip agency from those within? Or is it merely convenient? In Colnbrook Immigration Removal

---

[5] I am grateful to Azadeh Sarjoughian for her guidance on Persian art history.

Centre, for instance, we were told that the centre manager received leftover flowers from a florist. Yet such banal reasons are the private face of these institutions. Instead, positive stories are repeated about how a former detainee used his skills learned next to the runway to run a flower store in Bangladesh, after his deportation.

Still other interpretations may be possible. Flowers signal an alternative possibility, however out of reach it may feel, in these concrete and steel structures. Flowers seem to stretch into the sunlight available in enclosures, prisons, concrete jungles where delicate beauty is not presumed to survive. They are not necessarily thriving. But their presence in the men's artwork attests to that geometrically resonant beauty that wells up before us. They remind us of Uriel Orlow's work about political dissidents, such as Nelson Mandela planting a garden during his time on Robben Island,[6] or Edmund Clark's portraits of plants collected inside Grendon Prison, *In Place of Hate*.[7] What has brought the prison and the garden historically and globally into a commonality has to do with the liberty of growth that we attribute to the plant and yet paradoxically also its limited capacity, as we perceive it, to escape its habitat. The impossibility of self-uprooting and movement produces a kind of incredible survival in the limited confines of the plant body. In turn, the prison represents forced confines and enclosed non-growth, the opposite of the garden to the political prisoner, a curtailing of growth and a constraint on the proliferation of ideas in the world.

However utopian our attribution of freedom to plants is, the practice of drawing flowers in detention centres offers an ambivalent image of hope. For the otherness of plants can provide a sense of growth despite the control, restriction, and enclosure.[8] Botanical proliferation and change is indeed surprising in the context of human control and the social destruction of environments, and the delicacy of botany in even the most punitive, carceral, and colonial environments can become a source of

---

[6] On the prison and the garden in Uriel Orlow's work, see von Zinnenburg Carroll, K. (2018). 'Theatrum botanicum: restitutions to Nature's ghosts', in Sheikh, S. (ed.), *Theatrum Botanicum*. Berlin: Sternberg Press, pp. 237–43.
[7] Clark, E. (2017). *My Shadow's Reflection*. Birmingham: Ikon Gallery.
[8] See von Zinnenburg Carroll, K. (ed.) (2017). *Botanical Drift: Protagonists of the Invasive Herbarium*. Berlin: Sternberg Press.

great hope for humans. In the flower drawings from Colnbrook and Campsfield House it is not a question of the identification of species, nor of skill in their representation. Whether they are a first or a last resort, in the final analysis a flower represents a reduced paradise, which is still dreamt of, however skeletal its form.

The historical newspaper images of Harmondsworth remind us of the role of border control in managing decolonisation and the free movement of former British colonial subjects. Such matters were present from the very beginning and helped usher these institutions into law. As the women and men stand at the telephone or wait in the courtyards, a kind of hopelessness is apparent in their gestures. They are layered over the self-harm reports produced each month in every centre, which are rife with cases of attempted suicide. Such matters take a darkly comical turn into non-sense in the play, evident in the 'Depression Capacitar' (a self-administered 'therapeutic' approach advertised in the corridor of the centre behind Heathrow as a way to deal with depression and suicidal thoughts) as a contrast to the rhetoric of the Chief Executive Officer (CEO). The CEO of the fictitious Bordr Management Company justifies the spending cuts in terms of self-healing, improvement, and motivation. The petty nature of bureaucratic rule, the way in which denial is rendered into policy, contrasts with the embodied despair, then and now, of those waiting.

The guard drawn by a Somali detainee is being embraced lustily by a dog whose erect penis and open paws betray a kind of mad hope that the small unfinished flower in the bottom left of this drawing echoed, while the militarily clad figure of authority holds but promises that this relationship will remain paternal, and potentially punishing. The archive is full of adjacencies and juxtapositions that produce sinister associations. The drawing of the resplendent turkey shares a page with a portrait of a man and a gun that is larger than the man's face. It seems a memory of a person or self-portrait was often swiftly associated with the memory of their death, perhaps at the hands of the bearer of that gun.

The play traces the movement through the immigration removal centre and a day in the life of the building, which sees everything and is therefore able to voice a composite portrait from its own perspective. It does so in an electronic voice that further estranges its personality,

FIGURE 3.4 Khadija von Zinnenburg Carroll as CEO of BORDR in
Darwin College Lecture by Khadija von Zinnenburg Carroll, Cambridge, February 2018.
Photograph by Sircam.

FIGURE 3.5 *Men in Waiting.* Performance lecture by Khadija von Zinnenburg Carroll.

creating an ominous inhumanity and embodiment of the state. My voice and the electronic voice work together in contrast (alternating between scenes as set out in the script below).

Scene 1 begins with a poem that is a homage to Kamau Brathwaite, inspired by a review of my book *Art in the Time of Colony* that Richard Drayton wrote for *Third Text.*[9] Brathwaite's poem *Negus*, cited by Drayton, stayed with me, the repetitive negation of what he calls 'the semi-colony', which is expressed not as a lack of gratitude but as a call to attention: that the crossing of the border in itself is not enough, since it does not constitute the arrival of the migrant at a destination. What the detention centres show is that merely existing in a place is often hardly an existence at all.

By Scene 5, the play is in the courtyard of the centre, having moved through the rooms, turning from space to the time spent in it. Time is

---

[9] Drayton, R. (2015). 'Art in the time of colony', *Third Text*, www.thirdtext.org/art-time-colony (accessed 10 January 2019).

delineated around the 'hard lock up' during the night, which many detainees attested was the most difficult, isolating time. Scenes 7 and 8 are about the ways in which anxiety is translated into waking nightmares. Among the documents Mary Bosworth collected for the archive was an advertisement for the detention centre's high-grade 'Type E Luminaire' lighting. Christoph Balzar overlaid this with part of the poem from Scene 7 where the large puppet I made with many faces in one figure represents the Multitude of detainees. These prison lights that are 'vandal-proof' and 'tamper-proof' are significant also in Nana Varveropoulou's *No Man's Land*, in which she includes a photograph by a detainee taken at night of lights inside Colnbrook covered by a pullover. He said he used the pullover to block out as much light as possible, because his roommate couldn't sleep and had his light on (in the double cells the already-rife insomnia is heightened due to the sense of confinement and lack of darkness).

The slide-projection piece in Scene 5 counts the amount of time that it takes to walk around the courtyard smoking a cigarette. It plays on an old carousel slide projector, so, while the slides roll out in a circle, they mirror the walking in a circle that is shown on the stills. One circle follows another and is replaced by another lap in those days where there are two-hour slots for recreation. This is the timetable inside immigration detention: 9:45pm 'Quiet Time/Zone Down/Roll Call' lock up; 8am unlock 'Breakfast Starts'. There is no way of representing time like this, because it passes differently when locked up for 11 hours at night. Time changes shape. At night in a state of constant half-waking, time moves differently again. It becomes populated with nightmares, wild animals in survival struggles (there are many drawings of these: snakes strangling birds, fishes out of water, stampedes) full of fear and anxiety. The pictures are indices for these nights, drawings into darkness, of what has been seen in the mind's eye, and the shadowside.

As I perform this scene I walk in circles and notice that, focusing on my own breath, the space of the theatre is no longer the perimeter of vision. The immediate need for balance and breath blurs the distance to the walls. The deep space that is bounded by high walls (and a barbed-wire fence in the case of the courtyard) disappears as consciousness closes in on itself. A bounded enclosure circled by a body that wills its freedom from incarceration. That leaves no trace on the tarmac, whose last breath might be spent here or back in the horror we came from. The surveillance camera doesn't care either way about the breath that has turned to smoke. It cannot capture the burning inside the chest, the vibrations in limbs, the narrowing and deadening of the gaze.

There is, about the indeterminacy of detention that the criminologists have studied, a particularly detrimental effect on the psyche. If you are in prison you know how long you are in prison. But here I met people who had been here five years, who had returned three times over eight years, not to mention young adults whose whole future is being shaped by this waiting. They brought their family photographs and wanted to learn how to cut and paste themselves into another place in Photoshop, the computer software that makes it easy to extract your body from your surroundings and place it elsewhere. In this two-dimensional and potentially highly portable medium, migration could be enacted in its ideal state. The thin space of the digital photograph is a means of migration that also, in its flatness, accentuates the illusory nature of the image. A beaming selfie in front of a red bus, and other images of London as the icon of pulsing global capital.

Conclusion

As artists, we are asked to go into dark pockets that do not belong to the standard cut of the world. Pockets that have been grafted into the poor underbellies, the austere, strangled, and under-resourced jackets in which far too many pockets of belonging hang. It is into those that we are asked to shed some light, to grope about, to put our hands in and to feel what is there. What is the cost of this? It is certainly not remunerable. It is also made dubious in its scientific value because of the subjectivity of the artist who is enclosed by the darkness of this pocket, becomes part of that shadow world. There is both a necessity to enter and identify, be sympathetic, become part-of and a need to extract one's own making from the process of reflection.

FIGURE 3.6 Khadija von Zinnenburg Carroll and Multitude puppet in Scene 7 with a live video stream of a surveillance camera on the building (in the foreground) during rehearsal for the Darwin College Lecture, Cambridge, February 2018. Photograph by Sircam.

THE SCRIPT

Cast

BORDR Immigration Removal Centre, architectural model and LED light, 93 cm × 85 cm × 26 cm.

Imigrazie Puppets, leather, horn, starch, bamboo, paint, paper. *Multitude*, 86 cm × 115 cm, *CEO* 40 cm × 20 cm, *Protest* 29 cm × 16 cm, *I (Journalist)* 39 cm × 20 cm, *Building* 50 cm × 20 cm, *Mary Bosworth* 43 cm × 20 cm, *Van* 26 cm × 26 cm, *Hairy Angel* 24 cm × 43 cm.[10]

K., Khadija von Zinnenburg Carroll.

CEO of Bordr Management, resin mask, 21 cm × 14 cm.

English Lament, composition and live guitar by Jessyca Hutchens.

Anonymous Going in Circles, 80 colour slides.

Gandrung, musical composition by Mo'ong.

Redacted, The Secrets Act, overhead projection and drawings on transparencies.

Burn, digital video, 3:30 mins.

Scene 1

Total darkness. Puppet slowly emerges and breathes.

I   [I I]
It
it I
it is
It is not
it is not enough,
it is not enough to
it is not enough to be
to be here, to be paused, to be invisible, obsolescent
it is not enough to be silent, to be migrant

---

[10] The first version of this piece was made in Indonesia during a residency in Papermoon Puppet Theatre for their festival Pesta Boneka in 2016. The Indonesian version refers to the propaganda puppets inserted into *wayang kulit* by the Suharto regime to communicate to the multilingual masses. It was performed with sound by Johanes 'Mo'ong' Santoso Pribadi and translation by Yosephine Wastu Prajnaputri. The building was made with architect Lavinia Tarantino.

not in pure darkness   nor real sunlight
neither black   nor white
I am the shadow of whiteness, of lightening   whitening
and so the shadows   came to Cambridge.

The building's light comes on.

Scene 2

Building speaks in a robotic voice.

Welcome, my name is Bordr, I am an immigration detention centre in the United Kingdom of Great Britain. I will keep you here for the shortest time possible. As I am an Immigration Removal Centre the likelihood is you will be removed from the country when you leave the Centre.

I house up to 394 residents at any one time, all of which have different backgrounds, religious beliefs, and cultures but each and every one of them is treated with dignity and respect by the staff that work in me.

Scene 3

CEO monologue with mask on back of head facing audience. Slight movement like a puppet. PPT on screen.

Welcome to our newest site, I am the representative of our multinational organisation *BORDR Management,* you may have seen us around Cambridge before because we specialise in carpark security, and have expanded our portfolio to immigration centres where we deliver the best in controlled environments.

Ethics and an art of being ethical, a human touch, and that touch as we know we find in freedom we find in art work and it gives each and every human because everyone in here is an artist. We provide fantastic facilities, gifting residents with art supplies, space to work, recreation time, safety management.

The press has been spreading fake stories about us, you see how minor the deaths in our custody are, and they happen in the *gym!* of all great, healthy, beautiful places.

Really we are meeting today to celebrate our achievements.

These edifices represent the *care* that we take when anyone is in our custody. They are not just buildings, they are great cradling hands of care that protect the residents in our custody.

When we put our hands on any of the residents that have come from far and wide and who are hungry to learn and be part of this great nation and before they go home we provide them with further training so they can bring some of that great knowledge and humanity back to their own economies. All hands of our staff are equipped with the best new technology and devices for securing each and every body. In fact we contract the leading provider of Anti-Ligature Clothing – these are wearables not just to ensure the safety of those that travel alongside, from cell to van to charter flight home, but foremost to enhance comfort and protection, enable a relaxed position, ease of breathing.

We foster self-realisation, reliance: In our globalised world it is important to know where the borders lie. To know how doors are opened. That is in everyone's interests, and it is the creative energy and identity that Bordr Management, our company, represents.

Bends down and takes the mask off, turns to face the audience for the first time.

Scene 4

K. begins to circle around the whole room clockwise, to bring fire to the space through the clockwise pacing in circles, around the audience even, depending on the space. Include 'For Planning Purposes Only' here on a table or wall.

There is no address for my cul-de-sac at the back of the major airport.

Just a sign, which says *Care and Custody*. For those well versed in the euphemisms of the state, they know this must be the place. They drive in and wait in this car park.

I am, as it turns out, one waiting room after another. Unlike architectural programmes where each room has a different function, say the living or recreation room is one in which there is a carefree unstructured leisure time. The interesting thing about me is that these are all waiting rooms. The recreation rooms are actually waiting rooms, the dining room, the cell.

It is disorienting in there, labyrinthine. Instead of having enough staff to run me properly, they just build more doors. Heavy doors. Locks which I'm not allowed to speak about. The source of much emphasis: Do not photograph the doors. The official ban on the artist's documentation to begin with is on the locks and doors. Although actually they are magnetic so a photograph wouldn't be able to capture their particular power anyhow. Magnetic power. The difference between the magnetic power of lock opening and the kinds of holding of keys is a significant distinction between the so-called 'resident' of the waiting room and the guard of the waiting room. *[Accompanied by slides.]*

K. *turns to audience.*

How can a door in this world be cheaper than a person?
How *can* a door in this world be cheaper than a person?
How can a *door* in this world be cheaper than a *person?*

I was asked to describe what happened to a person here, and each time they go through the same rituals of waiting, of going through one locked door and waiting in front of another locked door. The experience of waiting by and looking through the small portholes in these locked doors became our relationship. In every door one could always see a face looking out with a mixture of mostly blank but partially still hopeful: Need. For it was the dependence on the guard with the keys that made this relationship so asymmetrical, so dependent, like a child on a parent. There were stories about female guards who locked in the inmates at night and then felt guilty – or perhaps it was desire – that drove them to call later on the telephone to talk to them through the walls.

Only telephones without cameras were allowed. A ban on images. But the men wanted particularly to have their image taken, their image given out, to be distributed, their image as affirmation that they are still here. But they are not here. There is no hereness in that space of waiting. And you cannot see them because all the images were destroyed. They are in that black hole of images into which all the evidence of this place goes. And yet we know from other crimes that those well documented are also those that can later be better retraced.

Think of extreme cases

because we have been here before   we have been here before

Kunst macht Frei, macht Kunst Frei? Frei macht Kunst

Capacitar wall recreated from Detention Centre photograph.

The paperwork is an aesthetic object which evidences the power of bureaucracy. In this Capacitar for curing your own mental health, read between the lines – pull yourselves together because there is no funding for anyone to help you – Now the way that these things are presented is in corporate cheer, the veneer of WordArt, which is freely available software that comes with Microsoft Word. The aesthetics of which are that of the pre-emoticon emotional richness, i.e. two-dimensional, garishly coloured array of kitsch icons with which a person is supposed to be able to identify as if beyond language, as if on a pseudo-emotional level of enthusiasm that is required in the corporate world, that is the unspoken contract between everyone that works underpaid, on a project they only superficially believe in, the ethics of which are mutually dubious to all, but which has such power of financial promise that in that WordArt is an expression of this emotional lie, of this complicity in a power structure that isn't equal. In a capitalist project that doesn't recognise the rights of the migrant worker or non-citizen prisoner, but that it acknowledges their 'aspirations' and 'well being' and 'capacitates' them in 'education' and 'recreation' and offers all these aspirational things that apparently the neoliberal world will enrich life with. WordArt is the aesthetics of neoliberal organisations, an anti-aesthetic that artists both despise and parody, as I am here.

I'm going to share with you now a part of a film we made, with material from inside the detention centres, as this whole piece is, from the art and media rooms within them.

'Artists in Residence' film is screened: https://vimeo.com/179157429

Fang became 'the most enthusiastic artist that Campsfield would ever have', according to the institution's newsletter. Profiled in interview within this document that is for internal circulation only, Fang was asked about how he sees his future with his artistic skills from arts school in China. 'I would like to do better in future to improve my current skill if I have the opportunity here in the UK' he replied.

Fang painted the Queen and Prince Charles, and the then Prime Minister David Cameron. He sent them their portraits with a letter setting out his case for British citizenship. Perceiving that horse portraits also appealed to the British ruling class, he painted horse heads for good measure. His works have the photorealism of painting from press photographs and the authenticity of a perfected mimicry, skills, we were told, acquired from working in China as a counterfeiter.

As a recent conference on *Faking, Forging, Counterfeiting* stated, although the visual impact of counterfeits contrasts with their negative connotation, they are still considered as fraud, fake, shadows of a creative act. There is a great irony, hence, in using photorealism to prove the artistry and authenticity necessary to be a skilled economic migrant. Yet forgeries can also be an 'embodiment of an aesthetic patriotism' and Fang's portraits can be read as interpretations of the press photographs on which he based his paintings. In his images, the establishment show the harsh public face presented to the press. It is the dignified official icon of the queen that Fang chooses to represent, not the paparazzi's scoop of her eating from plastic Tupperware in private. She is a decorated and upright figurehead, with the raised eyebrow of a judge in action. Like many others in detention, Fang assumed she has power in the UK, the way a hierarchical figurehead might. Of course, he was mistaken.

Fang's portraits are not uncontroversial in the detention centre. While the Queen sits proudly in the centre's boardroom, David Cameron has been returned, unwanted, to the art room. 'It is an election year,' we were informed, and like the BBC, if the centre were to hang a portrait of the incumbent, they would need to represent the other political parties as well.

The voiceover was text from of a series of posters I made during the Remain campaign, with Fang's letters in parafictional form. That is, the text was recreated from interviews about the letters Fang wrote to accompany his paintings (the originals remain with the recipients while the replies were treasured by Fang and taken with him when he was deported). In them the polite replies from Charles and the Queen are recreated to form a sequence. Apparently it all ended after Fang sent his portrait and letter to David Cameron. He was never seen again, deported once the gift arrived at number 10.

Scene 5

K. clicks through carousel slide projector of 80 slides while speaking the following.

1   Going in Circles

    people fly in circles
    collect points and classes, land   settle   plant roses
        my hands ache to touch the living
            let the sun hit my legs as I walk in circles
                let my feet feel the grass under my toes

breath, smoke, centre
     burn, body, decentre
mind flight
     hide me
        smoke screen centre
you hid me   London
      switched mind
      clarity building   day waning
lock making   time keeping
      today breaking   click click click
change   centre point to circle walk hold nowhere churn air camera
blind to   smoke in lung   churn body space
 in centre
          churn
    centre   churns body out of space
in the small rectangle cell   the circle skips corners
skips lives skips time
     circling record records lost.
           sounds dimmed. lights
  out. wake up.

2 WAKE UP!

Eat 9.
10. Legal.   11, 12, Fakery
           Recreation. Circles
Wait. Lunch. Lock up.   Roll Call
Held under.   Circling limbs in water.

I can't breathe. 4pm. Visits. I can't breathe. 4pm. You look me in the eye and I have to walk in circles for hours to forget how you look me in the eye. Lock me up for 12 hours and I cannot forget how you cannot look me in the eye. 12 hour circles.

Either the mind circles, small spastic trembling circles. Or my body in its massive machinic circling takes that little shivering mass and stretches it out. Over the tarmac.

You can't catch me so well if I'm walking. You can't pin me down nightmare, when I am running.

My shadow moves like the second hands of a clock
the clock moves like my legs   second   hour

we live by a clock that will not work for us
        does not change time   does not move forward
goes in circles as I go in circles
if only we could both stop
          face down circle
breathe out explode
        our arms could begin   to move backwards
unwind
      take uncertain steps into the future backwards
turn around and
stop   stop

3   Patterns Cast a Delicate Shadow

Beneath us the earth opens its mouth and breathes a sigh of relief
      spits out the concrete
           yawns and falls into a restful sleep.
For it has been rather stressful, to say the least, to have you all running in
  little blackened circles
        thick like oil   churning toxicity

the condition of movement   blown
on minimal redacted winds
black lines across white spaces
a dusty trace
      of parts that didn't survive
keep pushing me in circles
        all that will be left
is a black hole   for I will have become small enough to disappear
to trace your building with my body
      while I have to mirror your greed with my longing
        shrink my heart for your ego
keep walking keep saying there is a final solution?
        this can't go on
        this will always go on

Scene 6

The building speaks in a robotic voice.

I find intimacy with particular men in waiting by reading their emails,
stories of why they are here.

Watch their faces through the cameras: the face of 18-year-old Mohammed, whose fist fight one night over a girlfriend landed him in prison and now here, destined to return to an Afghanistan he fled from at age three, a place where there is nothing for him to return to.

They waited for the next opportunity to be allowed to continue.
To live.

Waiting in the twenty-first century could be accompanied by the Internet. Could be clicked through. Associatively. I can see what they see as they are waiting, for years. At least glimpses through what they search for online (which I monitor), images such as these (overfilled refugee boat *Vlora* leaving Tirana for Italy in 1992, which was showed to K. by a detainee as evidence of white refugees), and documents over email. It was the legal case that preoccupied them most. How to solve something that was mostly invisible, incomprehensible, always shifting, never accommodating.

Scene 7

'Multitude', a large puppet with many faces. Double flashlights, with Jessyca Hutchens as second puppeteer.[11]

I close my eyes, but it is never dark enough,
I cover my eyes,
your light is on the same switch as mine
It's on, because you can't sleep either
However much we cover our eyes

it is never dark enough

I see you there, sleep, but I cannot join you
I remain
without within
a world of surface
under which is hidden a shapeless mass

[11] This is the translation of a German version of this scene that was first written with the Swiss poet Jürg Halter for a performance of the play in the Konzert Theater Bern in March 2017. The puppet was made with Beni Sansaya and Yoghi Cahyo Nugroho in Papermoon Puppet Theatre.

the nights! – oh the nights!
the undark frightful nights
how the voices echo in that cell
walls, memories, hell
my love on the other side, of this hell
the undark frightful hell
of the nights, nights – nights!

Scene 8

Dream sequence of Hairy Angel puppet, with Jessyca Hutchens as second puppeteer of Mary Bosworth and CEO puppets.

I dreamt of walking through closed doors, dream that I become doors and walls, that I see myself in doors like in a mirror. I am a wall that tears itself down, and nothing but my shadow remains. I dreamt that I met the criminologist and the CEO at the border and we fought over who could fly.

Hairy Angel does a clock dance, swinging arms about to fight off the CEO.

I dream of flying, disappearing from here, dissolving, throwing my shadow on the other side of these walls, through the counterclockwise motion of clock-hands on a clock-face that runs clockwise towards death.

The guard's paperwork is always stacked high
and the alarm never stops ringing.

Scene 9

Building speaks and K. surveys it with a torch.

I see everything, you see, how they adulterate my food, add spices to chips, complain it's not the same rice as at home and run their own 'cultural kitchens'.

Cut open the wires of the electric kettles in the cells when they are locked up at night to make a spark, to light a cigarette. All of which is illegal.

Ach those stupid fake TV channels take these boy's stories about a few little rats that live in my garden, the garden that's even closed now, because no one could take care of it properly.

They write on me, how pitiful, all graffiti will be washed off tomorrow. Nothing sticks here. Yes, I have a radio station even, but it's not allowed to broadcast beyond my walls. I have newsletters, activities, it's a holiday in here. Okay so the walls smell of a fear you can never wash off.

Scene 10

Live redaction on overhead projector joins English Lament, *played by Jessyca Hutchens.*

Censorship began to play a part in the way we thought about what was possible, legally and artistically, what was necessary abstraction, and what is necessary documentation.

Redaction is said to protect the identities of those locked up, especially those who continue to challenge their asylum status and may, therefore, be vulnerable to the authorities in their country of origin.

Mostly there is this enormous language problem. It's not just the laws that seem to change, or at least remain so complex that they are not useful to the claimants. It is also their Englishness. It is the very Englishness, the very language. The very word that stands for not only for a language, English. But a way of being, English. A way of speaking, English. Belonging to the English. Being proper to the English. And those that are obviously un-English will quickly become evident through their lack of the command of English.

But you cannot be English. You cannot be a language. But you can be encased and closed and constituted by your use of this language and notion of being English, like a flag, waved vigorously, indicating, something.

Pause, CEO puppet waves his hand.

Murkiness. The images obscured entirely by murkiness, by censorship.

Rules and laws to be interpreted. Like language, except there is always a speaker. A guard who decides whether the interpretation was within the rules, within the law. To be outside of language, to be outside of the rules and law. That is where we find ourselves.

I am not only a place where people have dubious citizenship. I am an existential abyss that is far deeper than this broken nation.

Scene 11

Burn video projection, 3:30 min.

Note. The burning building at the end is a model that was built by a detainee in Colnbrook, which was destined to be thrown away before we included it in the archive in Oxford. When the Pitt Rivers Museum acquired that collection they didn't have space to store that large model, and again it was earmarked for destruction. Rather than see it thrown in a rubbish bin as an obsolescent artefact resonant thereby with the fate of the person who made it, I thought at least, since nobody could store it, it should be destroyed in an intentional way that also released and purified it in the process. One of the ways in which objects are ritually sacrificed is through fire, which is thought to also purify the material, polluted body or object, absolved through incineration. It was also an end that was desired, an end to a building that features as the central character. An end that is sung as it goes.

After the End

An audience member comes on set and picks up the mask and guitar. After singing songs through the mask, spontaneously, with vigour, he says the mask and instrument seemed an invitation. And so the play goes on through the objects that have gained a life of their own.

FIGURE 3.7 Khadija von Zinnenburg Carroll with Hairy Angel puppet in Scene 8 of the Darwin College Lecture by Khadija von Zinnenburg Carroll, Cambridge, February 2018. Photograph by Sircam.

4 Refugees and Migration

FILIPPO GRANDI

This chapter explores the intersection between refugee movements and the broader phenomenon of human mobility today. My starting point is the experience of two young refugee women, Mariam and Semira.[1] They were born and grew up in Eritrea – a country which they describe as beautiful, but where 'security and hope are scarce for most people'. In Eritrea, military service is compulsory, and in practice, for many, extends for several years. Mariam and Semira, like tens of thousands of young Eritreans before them, fled the country after prolonged periods of service, their requests for demobilisation denied. They embarked on a dangerous journey through Sudan, where they first met, and across the border to Libya, travelling at high speed on pick-up trucks for a month through the harsh desert, watching others fall off and die of injuries or thirst, themselves surviving on biscuits and water.

Once inside Libya, they were put into what they describe as a 'people store' – a holding pen run by an armed group, where people were squeezed into small spaces on top of one another, in appalling conditions with no hygiene facilities, with no air or natural light for months on end. Together with others held there, they were repeatedly tortured so they would seek money for their release from their families. Even those able to raise the money, from their families or through charity at home, were transferred to other groups, where the process of ransom would start again. The two women were kidnapped and sold between three groups of smugglers during their time in Libya.

After some time, Mariam, Semira, and six other women were removed from their 'people store' and taken to Misrata, on the Libyan coast. Here,

[1] Names changed for confidentiality reasons.

79

they thought they might finally make it across the sea to Europe. Instead, they were separated from the other women and, for two weeks, locked in a room together and repeatedly raped – their rapists wearing masks, so it was impossible to know what they looked like or even how many they were. When it came to an end, they were once more crammed into another 'people store', but eventually managed to escape through the window of a filthy latrine. Overcoming their terror, they sought refuge in a mosque, where they were handed over to the police, arrested and transferred to a government detention centre in Tripoli. There, they found themselves again in indefinite captivity, in a deeply fractured country affected by widespread insecurity, in the hands of a government with limited authority and capacity to exercise its responsibilities.

It is a horrific story, but regrettably one that reflects the experiences of tens of thousands of women, men, and children who make the journey through the Sahel region and North Africa towards Europe every year. Some are refugees, driven by violence in Somalia and other countries blighted by deep-rooted conflict, or by repression and persecution, moving in search of safety and a solution to their plight. Others are migrants, drawn by the prospects of greater opportunity, their decisions nonetheless also in many instances shaped by a complex range of 'push' factors.

But regardless of how we categorise people on the move through Libya, the same questions present themselves. How, in a world of modern nation states, shared prosperity, and boundless capabilities, can people today still find themselves repeatedly exposed to the horrors of kidnapping for ransom, of imprisonment, of torture and rape?

What is wrong with our modern system of international cooperation, so painstakingly built from the ashes of the Second World War in pursuit of peace, security, human rights, and development, that it is so badly failing so many of those who are moving in search of safety, stability, and opportunity across the globe? And – most importantly – how can and should the world respond?

These are timely and important questions that are – rightly – high on the global agenda. In September 2016, the United Nations General Assembly adopted the New York Declaration on refugees and migrants – a series of political commitments to which Heads of State around the world signed up, aimed at delivering a more adequate and predictable response to large-scale population flows.[2] It included a new comprehensive framework for responding to major refugee situations that has since been applied in 14 countries and two regional situations, and is already generating important results. Among other things, the Declaration called for the development of two global compacts – one on safe, orderly, and regular migration, and one on refugees.

Refugees and Migrants

In 1950, when the Office of the United Nations High Commissioner for Refugees (UNHCR) was established, the distinction between refugees and migrants was fairly clear. The first High Commissioner, my predecessor Gerrit Jan van Heuven Goedhart, was a Dutch journalist and politician who had been active in the resistance during the Second World War. Describing the dozen or so million people who were left scattered across Europe, far from their homes and countries at the end of the war, he noted that this 'was not a movement of people, trekking with their families and chattels from one destination to another, but the movement of uprooted individuals'.[3]

By 1953, around seven million had been repatriated, and one million had found homes in new countries. In surveying the task ahead, he noted that resettlement opportunities for the remaining refugees were narrowing, and set out proposals for their assimilation in the communities in Europe where they were living, with a particular focus on economic integration – a successful project that led to UNHCR being awarded its first Nobel Peace Prize the following year.

Fast-forwarding to 2018, the categorisations of the post-war years are still relevant, but sometimes less evident. The essence remains the same. Refugees do not move out of choice – they are forced from their countries in search of safety, owing to conflict, violence, and persecution. When more than half a million ethnic Rohingya fled from Myanmar to

---

[2] New York Declaration for Refugees and Migrants; Resolution adopted by the General Assembly on 19 September 2016 A/RES/71/1.

[3] See van Heuven Goedhart, G. J. (1953). 'People adrift'. *J. International Affairs*, 7(1), 7–29.

Bangladesh in the space of a month in September 2017, there could be no illusion that they were moving voluntarily, in search of a better life. The same applies to the 2.4 million South Sudanese who have fled vicious fighting in their country over the last five years, or the 6.3 million Syrian refugees scattered across the Middle East and beyond.[4] The flow of refugees, alas, is not 'orderly'. Fleeing for their lives, their movement is often chaotic and improvised, their assets left behind, and they are rarely able to cross borders through regular immigration procedures. Tens of millions more remain internally displaced within their own countries, often trapped and unable to flee abroad.

For most of the world's migrants, on the other hand, it is hope, rather than despair that motivates their decision to move. The vast majority of the 247 million people living outside their country of birth today have moved through regular migration channels, their decisions shaped by migration policies and labour demand in destination countries rather than by war, repression, and violence at home. And while the absolute number of international migrants is higher than ever before, as a proportion of the world's population, at 3.4%, it has increased only marginally since the 1960s. In short, migration is an intrinsic aspect of social and economic development, not a problem to be solved.[5]

And yet, the elements that allow conflict and persecution to take root and flourish often overlap with factors that shape migration flows: weak governance, impoverishment, deep inequalities, environmental degradation, lack of resources, water scarcity, and food insecurity. Climate change is an increasingly prominent driver of many of these trends. And in a world in which greater overall prosperity has been accompanied by deepening inequality (as documented by Thomas Piketty and others in the 2018 World Inequality Report), it is precisely those most affected by

---

[4] As at 31 December 2018, there were an estimated 68.5 million people forcibly displaced worldwide, including 25.4 million refugees. See UNHCR (2017). *Global Trends: Forced Displacement in 2017*, www.unhcr.org/globaltrends2017/ (accessed 8 October 2019).

[5] McKinsey Global Institute (2016). *People on the Move: Global Migration's Impact and Opportunity*, www.mckinsey.com/~/media/McKinsey/Featured%20Insights/ Employment%20and%20Growth/Global%20migrations%20impact%20and%20 opportunity/MGI-People-on-the-Move-Full-report.ashx (accessed 8 October 2019).

lack of development who, even if they can find the means to move, are likely to be excluded from legal channels and compelled to embark on irregular migration pathways.[6] In addition, for those countries that are transitioning out of conflict, and may have in the past generated refugee outflows, human mobility may emerge as an important coping mechanism, linked to poor governance, collapsed economies, and a lack of jobs and services.

The phenomenon of 'mixed' migratory flows has also become increasingly prominent, with refugees and migrants often moving together across spaces on land and at sea in which power is exercised not by the state, but by parallel forms of authority that have filled governance vacuums, including transnational armed groups, and smuggling and trafficking networks. The logic of deterrence and control that now prevails in many regions, insufficient regular migration pathways, and the absence of solutions for the overwhelming majority of today's refugees are together propelling both refugees and migrants into exploitative situations and dangerous journeys across mountains and seas.

The refugee dimension of these flows varies. In some cases – for example, the surge in arrivals in Europe by boat in the Eastern Mediterranean in 2015 – the vast majority are refugees. In others, the composition is more mixed. The current movements through Sub-Saharan Africa into Libya, on which Mariam and Semira embarked, consist primarily of migrants, with a smaller but significant refugee component.

Yet, once on these journeys, the risks to which refugees and migrants are exposed are often the same, as we are repeatedly and starkly reminded by the relentless toll of deaths at sea, by the accounts of people freezing to death as they cross high mountain passes, and by the chilling stories of brutality and exploitation by traffickers shared by Mariam, Samira, and others. The growing number of unaccompanied and separated children making these journeys is of deep concern – a phenomenon which UNHCR has documented in a number of regions over the last few years, from northern Central America, to the overland route taken by young Afghans across Central Asia to Europe, to the Eastern and Central

---

[6] World Inequality Lab (2017). *World Inequality Report 2018*, https://wir2018.wid .world/files/download/wir2018-summary-english.pdf (accessed 8 October 2019).

Mediterranean routes. Of the 119,000 people who arrived on Italy's shores in 2017, 13% were children unaccompanied by an adult or separated from their families.[7]

Refugees as a Distinct Group

Despite the complex dynamics driving today's population flows, and the common dangers to which so many of those on the move are exposed, it remains critical to avoid blurring the line between refugees and migrants. Refugees move because of a failure of protection in their own countries. They have fled war, violence, and persecution. Many have been displaced inside their own countries first, or have been trapped in enclaves or besieged areas, sometimes for years. Returning home, without a resolution of the conflict or repression that drove them from their countries in the first place, is not an option. Deprived of the protection of their own governments, refugees are considered to be of international concern, and the obligation to provide international protection lies at the heart of the 1951 Refugee Convention, for which UNHCR exercises a supervisory role, as well as its 1967 Protocol and later regional instruments.[8]

It is important to understand the background to this framework. The first mechanisms for international cooperation to address the situation of refugees were established by the League of Nations in the interwar years. International protection was extended to specific groups, including Russians, Armenians, and Assyrians, who were outside their states of origin and were politically excluded. Since they were impoverished and denied the possibility of returning home, the introduction of the Nansen passport in 1922 enabled many to move onwards to join family and pursue labour opportunities in new countries, including through schemes administered by the International Labour Office. Nansen passports were eventually recognised by 52 governments. Facilitating refugees' freedom of movement operated as a form of international burden-sharing, and

---

[7] UNHCR (2017). *Italy Country Operational Update – December 2017*, https://data2 .unhcr.org/en/documents/details/61549 (accessed 8 October 2019).

[8] 1951 Convention relating to the Status of Refugees and 1967 Protocol Relating to the Status of Refugees; 1969 OAU Convention Governing the Specific Aspects of Refugee Problems in Africa; and 1984 Cartagena Declaration on Refugees.

migration became part of the solution.[9] But international protection was the cornerstone on which those solutions were ultimately built.

By the 1930s, as the impact of the Great Depression hit, both asylum and migration frameworks had effectively collapsed. Without international cooperation to ensure protection, increasingly restrictive, state-by-state immigration policies proved a devastatingly inadequate response to the tragedy of Nazi Germany. Solidarity failed, and as a result, millions perished, because they were unable to flee the Third Reich, or were not given asylum, leaving a stain on the world's collective conscience that we would be well advised to recall today – when the very principle of international refugee protection is sometimes called into question, including by some of its traditional champions in Europe.

It was this experience that drove the United Nations, in one of its earliest acts, to adopt a resolution recognising refugees as a matter of international concern in December 1946, and to the international protection mandate entrusted to UNHCR just a few years later.

The inability of refugees to return home because of conflict or persecution, and the failure of national protection, are at the heart of their specific status, and the internationally agreed framework of rights and obligations that applies. Treating them as a sub-group of irregular migrants risks obscuring their distinct status and rights in international law, and can give rise to an approach in which control takes precedence over protection. On account of their coming from countries affected by conflict, violence, and, for some, violent extremism and terrorism, failure to pay attention to the causes of their flight, and to the reasons why they have been forced into travelling through 'illegal' means, also risks their being perceived as a potential security threat, rather than people who have fled insecurity in search of refuge.

It is also important to recall that the vast majority of refugees are hosted in countries neighbouring their own, and don't move further afield. At the end of 2017, the number of refugees worldwide stood at some 25.3 million, of whom approximately 85% were hosted in countries in developing regions, which included nine of the 10 largest refugee-hosting countries.

---

[9] Long, K. (2013). 'When refugees stopped being migrants: movement, labour and humanitarian protection'. *Migration Studies*, 1(1), 4–26.

Compare, for example, the one million refugees who in 2015 arrived in Europe, with a combined population of 500 million people, to the one million refugees who have fled from South Sudan to just one country, Uganda, with a local population of just 43 million. The vast majority of refugees are not 'on the move' but struggling to survive in countries bordering their own, trying to rebuild their lives in anticipation of the day when they will – one hopes – be able to return home.

With effective conflict resolution increasingly a rarity, for many that day becomes more and more distant. Around the world today, tens of thousands of refugees are weighing the difficult choice between staying in tough circumstances in exile or returning home to a still-risky situation, often amidst widespread destruction.

When protection fails, or the prospect of a solution back home fades, the decision to move in search of greater protection and stability becomes more likely. The movement of Syrian refugees towards Europe coincided with a moment when food rations in Lebanon and Jordan were being cut, only half of refugee children were in primary school, and the war in Syria had entered a new and devastating phase. It is also no coincidence that two of the world's 'global' refugee populations, frequently to be found in 'mixed' migratory movements, are Somalis and Afghans – both from fragile countries where opportunities remain few.

Turning to Mariam and Semira – in choosing not to remain in the refugee camps in eastern Sudan, they would have been profoundly aware of the bleak prospects of a life stretching ahead there, confined to a camp that was established up to 50 years ago, with no prospect of being able to work or build a future, and exposed to the many risks associated with long-term confinement, frustration, and lack of hope. They may also have been aware of a string of incidents of abductions and disappearances of Eritrean refugees there, allegedly involving border tribes, held for ransom or trafficked onwards for the purpose or forced marriage, sexual exploitation, or bonded labour, as well as reports of attacks and deportations.[10]

[10] See, for example, UNHCR (2013). 'UNHCR concern at refugee kidnappings, disappearances in eastern Sudan', www.unhcr.org/news/briefing/2013/1/5102 75a19/unhcr-concern-refugee-kidnappings-disappearances-eastern-sudan.html; UNHCR (2015). 'UNHCR deeply concerned about abduction of asylum-seekers

Approaching refugees through an 'immigration' lens obscures this broader context and the reasons that force refugees to flee their countries, or to move onwards from precarious situations in first countries of asylum. It allows irresponsible political leaders to exploit genuine public concerns generated by haphazard and reactive responses to refugee flows. And the practice of branding refugees as 'illegal immigrants' or otherwise avoiding the 'refugee' terminology, and the rights and obligations that it implies, is one which some countries hosting large numbers of refugees have closely watched, if not emulated.

I have taken some time to set out the background to today's refugee flows, and why it is important that the international refugee protection regime remains at the forefront of the international response. Yet, as we have already seen, there is an important intersection between refugee protection and migration, especially in 'mixed' migratory flows: the causes are often intertwined, they may be exposed to the same appalling risks en route, and as they progress towards their eventual destinations face the same consequences of xenophobia, nationalist senti-ment, and dehumanisation. Addressing the situation of people on the move, and protecting their rights and welfare, calls for a range of responses, some common to all, some differentiated according to status and protection needs.

International Response

What, then, are the fundamental considerations that should shape our response to all people on the move, and especially those travelling in today's 'mixed' migratory flows?

First and foremost, protecting the lives and dignity of all must be at the centre of the response. Rescue at sea, for example, is a humanitarian imperative, as well as a binding obligation. In 2015, UNHCR, the Inter-national Maritime Organization, and the International Chamber of

in eastern Sudan', www.unhcr.org/news/press/2015/6/5571bce09/unhcr-deeply-concerned-abduction-asylum-seekers-eastern-sudan.html; and Radio Dabanga (2017). '22 Eritrean refugees freed by traffickers in Sudan', https://reliefweb.int/report/sudan/22-eritrean-refugees-freed-traffickers-sudan (all accessed 8 October 2019).

Shipping issued guidelines for conducting search and rescue operations in line with international legal standards. But, for this to be truly effective, states need to share responsibility for deploying search and rescue operations and disembarking those rescued – either on a temporary basis, until they can be evacuated onwards or helped to return home, or by admitting them into their own asylum procedures.

Alternatively, and especially in exceptional circumstances when mixed migratory routes converge – as, for example, in the case of the Mediterranean – regional mechanisms to relocate those disembarked could significantly alleviate the pressure on those countries most affected. In 2017, for example, more than two-thirds of the 172,000 people arriving by sea in Europe were disembarked in Italy, with the remainder split more or less equally between Greece and Spain. As in 2015, when the bulk of the arrivals reached Germany, Sweden, and Austria, European solidarity failed to materialise, leaving a few countries to bear the brunt of the situation.

Second, misconceptions and myths around refugees and migrants must be challenged and countered. This must be done more vigorously, by governments and leaders that often appear frightened by populist rhetoric. It must also rest on evidence-based reporting and policy-making.

Let me mention a few of those misconceptions. I have already highlighted that, while migration is increasing in absolute terms, the number of those living outside their own countries as a proportion of the global population has remained at more or less the same level (around 3%) for the last 60 years. Around half of migrants globally have moved from developing to developed regions, but there is also a sizeable South–South movement, with one-third of the world's migrants having moved from one developing country to another.[11]

Similarly, while the proliferation of armed conflict that we observe today has resulted in a massive upsurge in forced displacement, and the persistence of perpetual refugee situations that show little prospect of being resolved, the impact is overwhelmingly absorbed by the developing world. Of the 68.5 million refugees and displaced people around the world today, more than 40 million are internally displaced within the borders of

---

[11] McKinsey Global Institute (2016), *op. cit.*

their own countries – unable to leave owing to increasingly restrictive border policies or because the brutal way in which today's wars are waged, with little regard for the lives of civilians, means that their path to safety is blocked.

The number of refugees worldwide has indeed now exceeded its previous peak in the 1990s – a deeply worrying development that is linked to the rise in the number and severity of conflicts over the current decade, and the inability of the international community to resolve them. Yet, as an overall percentage of the world's population, it is still somewhat lower. The desperate situation of the world's refugees, and the pressures on the countries hosting them, compel international attention and support. But this is achievable; it is not a matter that is beyond our shared capabilities – provided that states collectively step up to the challenge.

Migration and refugee flows are therefore not 'out of control' as some politicians would like us to believe. However, poor management, reactive, improvised, and piecemeal responses, and inadequate integration measures can give the impression that this is the case. As we have seen in Europe and elsewhere over the last few years, this can foster genuine fears regarding jobs, identity, and security – which are easily manipulated to allow xenophobia and racism to flourish – including, and sometimes especially, in those countries that have in reality received the smallest numbers.

Third, responses to mixed migratory movements must be based on a real understanding of what is driving and shaping these flows – and a comprehensive response that engages with these elements, in all their complexity. This means listening – to refugees and migrants themselves, to the governments and communities in countries that host refugees and through which they transit – and understanding the local political economies that allow trafficking, smuggling, and the small-arms trade to flourish.

An imbalanced emphasis on closed borders, containment, and deterrence as the response to the abuses perpetrated by traffickers and smugglers is not the answer. It simply drives refugees and migrants further into the hands of those who seek to exploit and abuse them. More opportunities for people to move regularly, including through migration schemes that meet labour market needs, and resettlement and other protection pathways for refugees will translate into better management of migration and refugee flows, as well as safe alternatives to perilous journeys.

Certainly, strong, collective action is needed to tackle the horrific abuses perpetrated by traffickers and to identify and prosecute them. Important initiatives have been undertaken by the United Nations Office on Drugs and Crime, Europol, the European Union Naval Force, the International Organization for Migration, and others. UNHCR has also made specific recommendations to tackle trafficking, including by freezing assets, travel bans, disrupting the supply of revenues and materials, robust prosecutions and sanctions against known senior figures and companies engaged in trafficking.

And, as we have seen in the Agadez region of Niger, hard security measures can create instability if not matched with adequate development investments that provide a real alternative to the smuggling industry – which is estimated to have offered direct jobs for more than 6,000 people, and to have indirectly benefited more than half of all households in Agadez. They may also expose refugees and migrants to greater danger as the cost of journeys goes up and the risk of being abandoned in the desert is heightened, with more vehicle breakdowns as smugglers take less-travelled routes to evade law enforcement.[12]

For refugees, fleeing for their lives, a comprehensive approach is needed that addresses conditions in countries of origin, in neighbouring countries, and in transit. Strengthening refugee protection and offering solutions along the routes – including in countries such as Niger and Libya – is key. For this to work, partnership – across regions and institutions – is critical.

I referred earlier to the comprehensive refugee response framework, which forms the basis of the global compact for refugees. This is built on the important principle of shared responsibility with refugee-hosting states through forms of support that move beyond humanitarian aid and early investment in solutions.

---

[12] Clingendael Institute (2017). 'Roadmap for sustainable migration management in Agadez', www.clingendael.org/publication/roadmap-sustainable-migration-management-agadez (accessed 9 October 2019).

The model is already being applied in 14 countries, and two regional situations, with a particular focus on shifting away from encampment, and fostering the inclusion of refugees in local communities and economies. Development investments are integral to the new model; the World Bank has been heavily involved and is rolling out important new financing instruments for refugee-hosting countries, and a number of bilateral development agencies are also engaged.[13] There is also an important role for development aid in supporting the drive for solutions in countries of origin, by addressing the causes of conflict and fragility.[14]

This 'comprehensive' model has enormous potential and, if properly resourced and applied, including in protracted situations, will help obviate the need for refugees to embark on dangerous onward journeys. The model also incorporates an important focus on pursuing political solutions and the resolution of conflict inside countries of origin, to enable refugees to return home. It also has an important emphasis on providing solutions beyond countries of first asylum – a point that I will return to in a moment. It is already being applied – with some success – to address the mixed migratory movements currently affecting Central America and Mexico, with the United States and Canada as cooperating actors.

Fourth, investments in strengthening protection and support for refugees and their hosts in host countries and along mixed migration routes must be matched by access to protection and humane treatment wherever they find themselves – including in destination countries. UNHCR has issued guidance on how to respond to mixed migration situations in a protection-sensitive manner – including on how best to facilitate entry and reception, using screening mechanisms and differentiated procedures to channel people into appropriate processes, so that asylum systems don't become overwhelmed.[15]

---

[13] The World Bank Group (2017). *Forcibly Displaced: Toward a Development Approach Supporting Refugees, the Internally Displaced, and Their Hosts*, http://documents.worldbank.org/curated/en/104161500277314152/pdf/117479-PUB-Date-6-1-2017-PUBLIC.pdf (accessed 9 October 2019).

[14] World Bank (2011). *World Development Report: Conflict, Security, and Development*, http://documents.worldbank.org/curated/en/806531468161369474/pdf/622550PUB0WDR0000public00BOX361476B.pdf (accessed 9 October 2019).

[15] UNHCR (2016). *The 10-Point Plan in Action*, www.unhcr.org/uk/the-10-point-plan-in-action.html (accessed 9 October 2019).

This guidance includes practical mechanisms for identifying and protecting refugees and migrants with specific needs, including unaccompanied and separated children, survivors of sexual and gender-based violence, people with disabilities, and victims of trafficking. Migrants as well as refugees may find themselves in vulnerable situations requiring protection and assistance – given the circumstances they encounter during their journeys or in destination countries, or because of their individual characteristics or traumatic experiences. Migrants may even find themselves trapped and exposed to danger when conflict breaks out in a country where they are residing – as it did, for example, in Libya in 2011, which at the time was estimated to host around 1.8 million migrant workers.

The human rights of all those in vulnerable situations need to be respected, and their immediate and specific needs met. In certain cases, people who do not fall within the scope of the refugee definition may therefore be granted permission to remain in the countries where they arrive for compassionate or practical reasons.

The 1951 Convention has proven to be a resilient and adaptable instrument, applicable to situations and forms of persecution not explicitly envisaged at the time of its drafting, such as those related to gender, age, and sexual orientation, and gang violence. Those fleeing armed conflict and other violent crises generally fall squarely within the Convention's scope, and, in some parts of the world, the Convention has been complemented and reinforced by regional instruments.

In addition, certain people who are outside their country of origin but who do not qualify as refugees may in certain circumstances also require international protection, on a short- or long-term basis, where their country is unable to protect them from serious harm. This might – for example – include people who have been displaced across an international border as a result of disasters, but are not refugees.

Arrangements for the return and readmission of people who do not qualify for international protection or meet other entry requirements are also an important part of a well-functioning migration management system, as well as a credible asylum process. Wherever possible, voluntary return should be pursued, accompanied by support for reintegration, and in all cases returns should be carried out in accordance with human rights standards.

Last – and most importantly – we must strive to restore a sense of solidarity. Over the years, there have been important examples in which energy, creativity, and political commitment have converged and states have collectively mobilised to address large-scale displacement crises through a range of instruments, including the transfer of refugees to other countries. States joined together to resettle Hungarian refugees in 1956, to respond to the Indochinese crisis (including through an innovative orderly departure scheme from Vietnam), and to address outflows from the former Yugoslavia in the 1990s.

Refugee resettlement has remained an important instrument, allowing refugees with particular protection needs, including women at risk and survivors of torture, to be resettled to third countries where they and their families are able to build a future. Yet, while the number of places available rose steadily over much of the current decade, and a growing number of states have been engaged, including the United Kingdom, the number of refugees resettled annually has remained well under 1% of the global total – a relatively insignificant number. Following a sharp reduction in the number of places made available in 2017 by the United States, traditionally a global leader on resettlement, the number of refugees able to benefit from UNHCR resettlement referrals dropped by more than 50% to 75,000 in 2017 – a significant setback.

In another example of how migration and refugee protection intersect, resettlement opportunities could and should be complemented by labour migration, family reunion, scholarships, and other migration avenues. This is not a new idea – migration schemes were a key aspect of international efforts in support of High Commissioner van Heuven Goedhart's work to resolve the situation of the millions who remained displaced in Europe in the early 1950s. A number of initiatives are already under way or in development. Although their scope currently remains limited, if they were scaled up, they could provide a powerful tool for international responsibility sharing in support of refugee-hosting states.[16]

[16] See Long, K., and Rosengaertner, S. (2016). *Protection through Mobility: Opening Labor and Study Migration Channels to Refugees.* Washington: Transatlantic Council on Migration, Migration Policy Institute, www.migrationpolicy.org/research/protection-through-mobility-opening-labor-and-study-migration-channels-refugees (accessed 9 October 2019).

There is, quite simply, no other way to address the phenomena of large-scale refugee flows and mixed migration, except through international cooperation. In the case of Libya, this is now being pursued through a joint African Union–European Union–United Nations Taskforce to Address the Migrant Situation in Libya set up in December 2017. In October, the International Organization for Migration stepped up its efforts to facilitate the voluntary return of migrants from Libya to their home countries, and by mid 2018 had helped several thousand return home. UNHCR secured the release of over 2,900 detained asylum seekers and refugees in 2017 and the first half of 2018, and, as of November 2017, embarked on a major effort to relocate asylum seekers and refugees out of detention centres, by transferring them to emergency evacuation facilities, particularly in Niger – an enormously complex, but potentially life-saving, exercise, involving the cooperation of a number of states.

Since they are unable to return home owing to conflict and persecution, the solutions for this group will of necessity be different from those for migrants, who form the vast majority of those detained. By July 2018, 1,858 men, women, and children had been evacuated, and more than 200 of those had already been resettled onwards to third countries from the evacuation facility in Niger.

If we are serious about addressing mixed migratory flows, resettlement and other pathways for admission to other countries must figure prominently as part of a comprehensive response.

Conclusion

When Mariam and Semira were in the government detention centre in Libya, they came to the attention of a UNHCR protection team, who negotiated their release and arranged for their evacuation from Libya to Niger in December 2017. They flew to Niamey along with 72 other refugees and asylum-seekers – mainly Somalis and Eritreans under the age of 18 – and were taken to a guesthouse there, supported by UNHCR.

When they were interviewed by UNHCR colleagues shortly after their arrival, they shared their relief at finally having escaped the traumatic situation that they had endured for so long. 'We just want to close the door on this experience and move on to a new and peaceful life,' Mariam said. 'We don't want much – just to be educated so that we can work to help our families. And we want to be safe.'

Her words, and the hope and expectations of those two young women, need no further comment. Simply put, we cannot let them down.

5 Migration of Disease

Dengue and Zika across Continents, around Cities, and within the Human Host

EVA HARRIS AND FAUSTO BUSTOS CARRILLO

Dengue virus, the cause of tens of millions of cases of dengue annually, and Zika virus, the cause of recent explosive epidemics across Latin America associated with congenital defects and microcephaly, are pathogens with a significant burden on human health and well-being. Beyond their toll on health, these flaviviruses, as the genus is called, share a common evolutionary origin, have each adapted to replicate efficiently in the human host, and are spread by similar species of mosquitoes. These commonalities belie important clinical, epidemiological, and immunological differences between them that necessitate continued research into their respective biology. Moreover, these commonalities and differences have real consequences for control efforts in affected areas around the world.

This chapter will explore the migration of dengue virus and Zika virus across continents, around cities, and within the human host. Viral evolution will also be framed as the continued migration of mutations across time and space. Together, migration and evolution will shed light on the origins and impact of these two viruses. These two concepts will also address existing questions that are critical for making progress in controlling these challenges to human health.

Dengue

Dengue virus (DENV) is widely regarded as the most important mosquito-borne viral disease of humans. Approximately 3.0–3.5 billion people are currently at risk of infection (Guzman *et al.*, 2010; Murray *et al.*, 2013). The virus is primarily transmitted by mosquitoes of the *Aedes* genus, especially *Aedes aegypti* and *Aedes albopictus*. These two *Aedes* species are found throughout tropical and subtropical regions of the planet. Female *Aedes* mosquitoes transmit the virus between humans when they take a blood meal.

Each year, up to an estimated 400 million new DENV infections occur, and of those, approximately 100 million can go on to develop clinical disease manifestations (Bhatt *et al.*, 2013). Dengue presents a global public health problem, and its continuing spread around the world poses significant challenges to control efforts. The virus has found new areas in which to invade, disseminate, and propagate, while global surveillance and reporting of dengue cases have improved over time, together resulting in dramatic increases in reported dengue incidence worldwide over the past several decades (Al Awaidy *et al.*, 2014; Guzman *et al.*, 2016; Wilder-Smith and Gubler, 2008).

DENV has four distinct serotypes – viral subspecies defined by their surface antigens (proteins on the virion surface that elicit an immune response) and genetic relatedness – unimaginatively called DENV-1, -2, -3, and -4. In 1975, and presumably before then, DENV serotypes were relatively isolated. With the exception of Southeast Asia, where all four serotypes co-circulated, other regions of the world had at most one or two serotypes (Guzman *et al.*, 2010). Since then, international trade and travel have disseminated the four serotypes across the world, and now the four serotypes co-circulate in almost all tropical and subtropical regions of the world. Interestingly, the greatest risk factor for severe disease is a previous infection with a different serotype (Halstead *et al.*, 1967; Sangkawibha *et al.*, 1984; Thein *et al.*, 1997). Thus, severe disease is thought to involve both immunopathogenesis (disease processes potentiated by the human immune response) deriving from the interaction between the immune response to a prior DENV infection and a current infection as well as viral pathogenesis (disease processes initiated by viral proteins). As a result, dengue is quite a complicated disease that has been the subject of much research over the past few decades.

Only about a quarter of persons who become infected go on to develop symptoms; the rest of the infections are inapparent or asymptomatic. However, symptomatic DENV infections can be severe. The clinical manifestations of classic dengue (also known as breakbone fever, or *la quebradora* in Spanish) include acute onset of very high fever as well as headache, retro-orbital pain (pain behind the eyes), muscle pain, joint pain, and rash. A secondary infection with a different serotype from the first increases the potential for haemorrhagic manifestations and vascular permeability, which leads to internal leakage of fluids out of the circulatory system and shock (Sangkawibha *et al.*, 1984; Thein *et al.*, 1997; World Health Organization, 1997). Dengue is a frightening disease from a population perspective because so many infections occur every year and because it is not easy to identify those who are at highest risk of severe complications. In fact, once vascular leakage begins, even previously healthy individuals can die within 24–48 hours. Oftentimes, the most severe clinical manifestations of dengue infection occur in children, leading to a high loss of life among the most vulnerable.

The dengue virion is covered with an envelope (E) protein that forms antiparallel dimers across its entire surface. This E protein mediates the infection process, whereby a virus enters a human cell and begins replicating. Because the E protein is on the surface of the virion, it is a main target for human antibodies, components of the immune system that seek out foreign material and target them for neutralisation or elimination (Lai *et al.*, 2008). The single-stranded viral RNA genome is found inside the virion. Because the cellular machinery that replicates RNA lacks a proofreading activity, RNA replication has a higher mutation rate than DNA replication. The higher mutation rate for RNA viruses like DENV and Zika (ZIKV) allows them to generate new genetic variants, select adaptive mutations, and quickly propagate those mutations to the next generation of viruses.

Genetic sequencing is the process of analysing and determining the sequence of nucleotides – the As, Us, Gs, and Cs – that constitute an RNA genome. As a result of evolutionary processes, sequencing and comparing the genomes of different organisms can reveal the genetic distance between them, a marker of how close different organisms are to each other on the tree of life. DENV-1, -2, -3, and -4 have very similar nucleotide sequences, indicating that the four DENV serotypes are, roughly speaking, genetic siblings (Barba-Spaeth *et al.*, 2016). Out of all the viruses that researchers have examined, ZIKV is the most closely related to the DENV serotypes on the family tree, but distinct in its own right – indicating that ZIKV is more akin to a genetic cousin of DENV.

Unfortunately, the underlying genetic similarity between DENV and ZIKV presents a significant problem for public health surveillance systems. Because DENV and ZIKV are so similar genetically and anti-genically, the antibodies that the human immune system generates against each virus will also be quite similar. As a result, many serological assays – tests that rely on antibodies to determine which virus has infected a patient – do not distinguish accurately between a current ZIKV infection and a prior DENV infection. This is due to antibody cross-reactivity, which has caused much confusion in laboratories around the world as well as with public health practitioners on the ground. We will revisit the topic and problem of antibody cross-reactivity in later sections.

The integral hypothesis of dengue is a general framework for consider-ing three levels of risk factors for dengue complications (Guzmán and Kourí, 2002; Kouri *et al.*, 1987). At the *individual* level, age, sex, nutri-tional status, co-morbidities, and immune status are important aspects of susceptibility that factor into a risk analysis. However, due to co-circulating DENV serotypes, there are additional *epidemiological* risk factors to consider. These include population-level immunity to each of the serotypes (the number of persons who are susceptible to each ser-otype), vector (mosquito) density, the spatial distribution of the mosquito vectors, and the number of circulating serotypes. Finally, *viral* risk factors must also be considered. These factors operate like a set of Russian dolls, since each serotype comprises a series of heterologous genotypes, each genotype contains a set of clades, and each clade is made up of a variety of strains. Additionally, it was recently discovered that a viral protein, NS1, can trigger vascular leakage directly and thus may play a role in the pathogenesis of severe disease (Beatty *et al.*, 2015; Modhiran *et al.*, 2015; Puerta-Guardo *et al.*, 2016). At each of these taxonomic levels, genetic differences can have an impact on disease severity. Entire scientific careers have been spent understanding each piece of the puzzle, but only by putting all the pieces together can we unravel the full, complex picture of dengue disease severity.

Zika

ZIKV was initially discovered in 1947 serendipitously when researchers identified a new virus in a single rhesus monkey in Uganda's Ziika Forest (Dick *et al.*, 1952). The first evidence of human infection was detected five years later in Uganda and the United Republic of Tanzania during a sero-survey of East Africa (Smithburn, 1952). The virus slowly spread across equatorial Africa throughout the next 15 years and later expanded into equatorial Asia, but no large outbreaks of disease were detected until after the turn of the millennium (Figure 5.1). Seroprevalence studies, which document population-level exposure to the virus on the basis of the antibody response, indicated that many people were becoming infected with ZIKV (Fagbami, 1979; Saluzzo *et al.*, 1982). However, very few cases of the disease were reported to public health authorities, possibly because Zika, dengue, and other mosquito-transmitted viruses have very similar symptoms. While the virus kept expanding its reach, it was never associated with severe disease and for decades was relegated to footnote status as an inconsequential virus.

This changed in 2007, when ZIKV caused its first large outbreak on the Pacific Island of Yap, part of the Federated States of Micronesia (Figure 5.1).

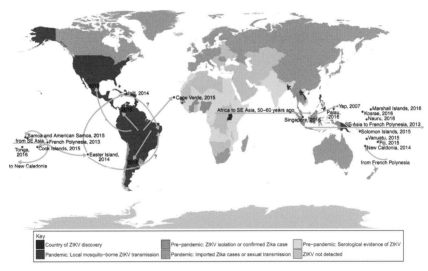

FIGURE 5.1 Map of the geographical extent of Zika virus dissemination. Figure by Fausto Bustos Carrillo. Data until 2 August 2018 are included.

An estimated 73% of the island's inhabitants were estimated to have been infected by the virus (Duffy *et al.*, 2009). From there, the virus slowly made its way across the Pacific. In 2013, ZIKV reached French Polynesia, where it caused another massive epidemic. About half of the island's general population was shown to have been infected in early reports, while approximately two-thirds of schoolchildren had been infected (Aubry *et al.*, 2017). The occurrence of a rare neurological condition called Guillain–Barré syndrome was linked to ZIKV infection during the French Polynesian epidemic (Cao-Lormeau *et al.*, 2014). Thereafter, the virus found its way into Haiti, then Brazil, and from there it spread explosively throughout the entire Latin American region (Faria *et al.*, 2017; Metsky *et al.*, 2017; Thézé *et al.*, 2018). The migration of the virus across these expansive geographical scales will be explored in the following sections.

Because so little was known about Zika disease at the time, it was very surprising when reports surfaced of ZIKV-infected mothers giving birth to babies with smaller-than-normal head circumferences (microcephaly). In adults and children, a typical case of Zika leads to a short-lasting rash that may be accompanied by mild fever, conjunctivitis (red eyes), joint pain, and/or muscle pain. A large proportion of asymptomatic or subclinical infections occur as well. However, babies born to mothers infected with ZIKV during pregnancy can display a range of alarming and likely irreversible symptoms termed congenital Zika syndrome (CZS). These symptoms include severe microcephaly with a partially collapsed skull, decreased brain matter, damage to the retina in the back of the eye, shortened and hardened joints (contractures), and muscular rigidity that limits movement (hypertonia) (Centers for Disease Control and Prevention, 2018). Brain asymmetry, absent or abnormal brain structures, and seizures are a few of the many symptoms that can present alongside CZS. The long-term consequences of CZS have not been fully studied as yet, but they might include permanent learning disabilities, abnormal motor-neuron development, and a high degree of morbidity over the life course.

Aedes **Mosquitoes**

The environmental commonality of DENV and ZIKV, as well as a vital link in the chain of transmission, is the mosquito. Although ZIKV can also be transmitted sexually, that route of transmission is considered secondary to mosquito transmission. The fundamental mosquito problem is the widespread geographical distribution of *Aedes* mosquitoes, especially the *Aedes aegypti* species. *Aedes aegypti*, the primary vector for DENV and ZIKV, is very common in most of Latin America, Southeast Asia, India, Oceania, and Sub-Saharan Africa, as well as the Gulf Coast and East Coast of the United States (Kraemer *et al.*, 2015). That *Aedes aegypti* can transmit diseases in so many areas of the world makes eliminating the species very difficult – let alone other mosquito species that also transmit diseases – by traditional means.

The emergence and re-emergence of DENV and ZIKV are strongly driven by global trade, global travel, and increasing urbanisation throughout much of the world. First, the ease with which modern shipping routes move cargo between and across continents has contributed to the introduction of pathogens and their vectors into new human populations. As an example, it is helpful to consider the case of *Aedes albopictus*, another mosquito of the *Aedes* genus that transmits DENV, ZIKV, chikungunya virus, and a variety of other disease-causing viruses. *Aedes albopictus* has a similar global distribution to *Aedes aegypti*, except that it is also found throughout the South, East Coast, and Great Lakes regions of the United States (Kraemer *et al.*, 2015). The mosquito's widespread distribution throughout the continental United States is thought to have originated from the importation of used tyres (likely from Japan or another northern Asian country) during the early 1980s (Hawley *et al.*, 1987; Kambhampati *et al.*, 1991). Trading routes shuttle used tyres throughout the world, and these tyres often contain enough water to host the developing and immature mosquito larvae. These infested used tyres are thought to have been the mechanism by which *Aedes albopictus* was introduced to the North American mainland, as the mosquito species was initially found in several tyre dumps in Harris County, Texas in 1985 (Sprenger and Wuithiranyagool, 1986). The larvae develop into pupae and then adult mosquitoes that lay eggs; the cycle continues, such that the new region becomes infested.

Second, global travel also functions in the same way: a traveller gets infected with a virus in one part of the world, doesn't know it, gets on a plane, and starts an epidemic when he or she returns back home and is bitten by a local mosquito that can transmit the virus. If this sounds like the plot of Richard Preston's *Hot Zone: A Terrifying True Story*, that's because the book's plot accurately characterises the emergence of exotic viruses as driven, in part, by humans moving a virus from place to place. In the case of mosquito-transmitted viruses like DENV and ZIKV, the mosquito vector must already be present at the traveller's destination for an epidemic to spread like wildfire through human populations. However, both global trade and urbanisation have resulted in mosquitoes being able to impact large human settlements.

Unplanned urbanisation in tropical areas often results in intermittent water supply or no water supply at all, especially in the poorest sectors of an urbanising city. In response, residents collect and store rainwater in barrels or other receptacles on their property. *Aedes* mosquitoes thrive in urban settings, since they breed in open containers with clean water. In a few days, the mosquito larvae develop into pupae and then adults, which fly around the house and a 50–100-metre radius in the neighbourhood, looking for blood meals and increasing the possibility of transmitting viruses between neighbours. This problem is compounded in areas that lack good waste management, as bottle caps, shoes, plastic lids, and most objects that are casually thrown away are perfect for catching rainwater and becoming mosquito breeding sites.

Owing to global trade, global travel, and continued urbanisation, mosquito and human populations are constantly interfacing. Diverse mosquito genera, including *Aedes*, are becoming increasingly adapted to the human ecology of the twenty-first century. Mosquito populations have been documented to adapt to insecticide use and human scents (McBride *et al.*, 2014). Some populations are even changing their biting behaviour to better adapt to human diurnal activity patterns (Russell *et al.*, 2011). Unfortunately, the more the mosquito becomes adapted to human habitats and behaviours, the more likely it becomes for mosquito-borne viruses to take advantage of this and spread much more efficiently through human populations.

Migration across Continents: Adaptation and Natural Selection

This section focuses on the migration of viruses across vast geographical spaces by considering the effects evolutionary adaption and natural selection have had on the viral genome. In *On the Origin of Species* (1859), Charles Darwin explained his notion of natural selection for adaptive traits this way:

> Every slight modification, which in the course of ages, chanced to arise, and which in any way favoured the individuals of any species, by better adapting them to their altered conditions, would tend to be preserved; and natural selection would thus have free scope for the work of improvement (Darwin, 1859).

At the time Darwin had this insight, he did not know that mutations, and genetics more broadly, provided the mechanism through which natural selection operates to select for adaptive traits. Evolution through natural selection is now regarded as a central and unifying concept across the life sciences, as it explains so many natural phenomena.

In this section and the subsequent ones, the evolution of DENV and ZIKV will be explored through the lens of migration. Specifically, evolution will be envisioned as a migration of mutations over time and space. The *geographical distribution* of DENV, as well as the migration of ZIKV across continents, can be explained by particular adaptive mutations that both viruses acquired over time. The *spatial distribution* of DENV and ZIKV infections – how these viral infections are embedded into the human ecology of a city – is very much a function of how well mosquitoes have adapted to an environment built by humans and the mosquito migration patterns therein. Finally, both DENV and ZIKV have acquired mutations that allow them to invade *different areas of the human host*. ZIKV, in particular, has an unusual capacity to invade across the placental barrier and affect a developing fetus. This literal migration of the virus through the human host is functionally responsible for the occurrence of CZS, which will be covered towards the end of this chapter.

Existing records suggest that DENV was introduced into the Americas in the seventeenth century by way of slave ships from Africa (Wilson and Chen, 2002). Historical accounts report dengue-like outbreaks throughout the Americas since then, including in the United States, especially along the Eastern Seaboard. Starting in 1947, the Pan American Health Organization, the World Health Organization's Regional Office for the Americas, began an intense campaign to reduce the *Aedes aegypti* population in North and South America to interrupt yellow fever transmission, as yellow fever virus is transmitted by the same mosquitoes as well (Pan American Health Organization, 1971). No dengue epidemics were reported from 1946 until 1963, reflecting the significant impact mosquito-elimination programmes had on human health throughout the continents. However, funding for the campaign dried up, and shortly after it stopped, most countries were re-infested with *Aedes aegypti*. Beginning in 1963–4 with a large dengue outbreak in Jamaica, DENV began to spread once more in the Americas (Ehrenkranz *et al.*, 1971). Within a few decades, dengue had been reported in 43 countries within the region and had become endemic throughout much of Central and South America. The disease remains endemic in the region to this day.

In contrast to historical DENV trends, epidemic ZIKV spread much more rapidly (Figure 5.2). After the French Polynesian epidemic, Easter Island in Chile experienced a Zika outbreak in February and March of 2014 (Tognarelli *et al.*, 2016). From Easter Island, ZIKV moved into Haiti in late 2014, leading to an outbreak (Lednicky *et al.*, 2016). In March of 2015, an international alert was sent from northern Brazil describing a mysterious, mild, dengue-like illness that was causing rash, low-grade fever, and generalised pain (ProMED, 2015). The disease spread quickly, so much so that 90% of doctors' visits in the initially affected area were due to this unidentified disease (ProMED, 2015). Within three months, a third of all Brazilian states were reporting similar outbreaks. In October 2015, five Colombian departments reported imported cases of Zika. By January of 2016, Zika was widely reported throughout Central and South America; imported cases were also reported in the United States, the European Union, the Middle East, and as far as Australia. In another two months, traveller-associated Zika cases would spread to Africa and Asia.

ZIKV's association with microcephaly and congenital birth defects in babies born to mothers infected during pregnancy was initially recognised in northeastern Brazil in 2015 and prompted the World Health Organization (WHO) to declare a Public Health Emergency in February 2016 (World Health Organization, 2016a). By the end of May 2016, a growing international consensus that ZIKV could be sexually transmitted – unusual for human flaviviruses – led the WHO to recommend safe sex or

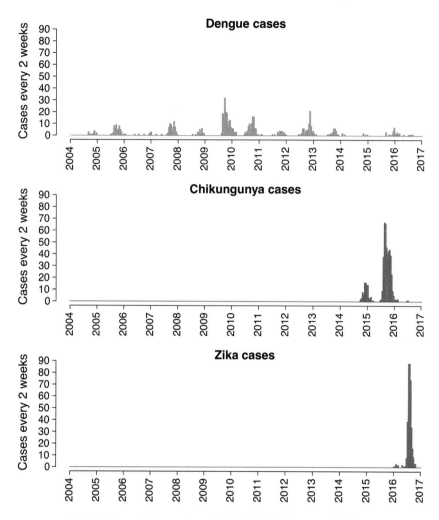

FIGURE 5.2 Time series and epidemiological curves for dengue, chikungunya, and Zika cases in the Pediatric Dengue Cohort Study in Managua, Nicaragua, 2004–17. Figure by Leah Katzelnick.

abstinence for eight weeks after travellers return from a place with reported Zika cases (World Health Organization, 2016b). Globally, the Zika pandemic has now calmed down. There are currently much fewer incident Zika cases in Latin America than in 2015 and 2016. In large part, this is due to a major portion of the Latin American population having been exposed and infected during the peak of the pandemic. As a result, it is thought that these previously infected individuals are likely protected against getting Zika disease in the future.

However, populations that were not widely exposed to ZIKV during the epidemic, such as some countries in Asia that are believed to have experienced only low-level transmission of ZIKV, are still at risk for a Zika epidemic in the future. Moreover, recent research suggests that Asian *Aedes aegypti* populations more easily transmit the pandemic ZIKV strain than ancestral, non-pandemic Asian ZIKV strains. This raises the worrying possibility of future introductions of pandemic ZIKV into Asia and subsequent epidemics throughout the densely populated continent (Pompon *et al.*, 2017). Although the WHO eventually declared Zika to no longer be a public health emergency (World Health Organization, 2016c), Zika is still viewed as an emergency within the scientific community because Zika clearly manifested pandemic potential and there is so much we still do not understand about the virus.

Molecular genetic techniques are a powerful suite of laboratory methodologies that are used to decode an organism's nucleic acid sequence. Scientists have used these methods to understand the history of ZIKV's dissemination and evolution, uncovering adaptive mutations that arose to make the pandemic strains of ZIKV behave differently from its ancestral strain.

DNA sequencing is the primary method used for the analysis of nucleic acid sequences. In the case of a single-stranded RNA virus like ZIKV or DENV, its RNA is first turned into a double-stranded DNA analogue by using the RNA strand as a template for the faithful conversion from RNA to DNA. This step is done because sequencing technologies are based on deciphering DNA instead of RNA. From there, the DNA in the sample is amplified (many copies of it are made) and sequenced (the order of different nucleotides is determined), and the resulting code is analysed by powerful bioinformatics software. Initially, these methods took all of the data from all of the DNA copies and averaged them to get a consensus sequence, a global representation of the most likely nucleotide at each individual site. Modern technologies produce the consensus sequence as well as the sequence of individual DNA strands in the sample. This allows scientists to understand the level of nucleotide variation that exists at individual sites within the genome and among individual DNA strands in the population.

For example, if a collection of similar viruses – say from several patients who displayed different symptoms during the same outbreak – were analysed, scientists would sequence the viruses. The genetic code of the different DNA strands could then be aligned and the strands compared with each other at every location in the genome to look for the presence of particular mutations that correlate with more severe symptoms. Imagine a scenario where 100 patients' samples were sequenced. If 18 of the 20 patients with the most life-threatening symptoms had a cytosine nucleotide (C) at position 134, whereas everyone else in the study had an adenosine (A), this difference would prompt scientists to examine the mutation more carefully. Every nucleotide triplet codes for an amino acid, and, strung together, amino acids form proteins, the molecular workhorses of every organism. An A-to-C mutation could code for a different amino acid, which in turn could alter the structure and/or function of a protein. And because RNA viruses have an elevated mutation rate, it would only take a small number of mutations in just the right positions for a virus to acquire greater lethality, transmissibility, or other adaptive properties.

Researchers have examined the geographical spread of ZIKV from Africa to Southeast Asia to the Pacific Islands to Latin America phylogenetically (Pettersson *et al.*, 2016). This means that they took viruses from different outbreaks that occurred at different times, sequenced the samples, and compared the viral sequences, looking for mutations that arose as the virus migrated from Southeast Asia to the Americas. In this manner, phylogenetic analyses reveal *when* and *where* mutations arose within the viral genome, and the underlying sequence reveals *what* each mutation was.

Two mutations that arose independently of each other have had specific functional consequences, one in the prM protein, the other in the NS1 protein, two of the 10 proteins encoded by flaviviruses. The prM protein is thought to be necessary for viral maturation and subsequent release from a cell (Pierson and Diamond, 2012). The occurrence of the prM mutation S139M coincided with the emergence of epidemic ZIKV with novel abilities to affect neural tissue (Yuan *et al.*, 2017). The introduction of this single mutation into the ZIKV genome was shown to result in significantly more microcephaly and higher neonatal mortality in mice than for the virus without the mutation. Scientists have also taken a pathogenic strain with this mutation, altered the virus to remove the mutation, and shown that mice infected with the altered virus develop normally. These two experiments provide strong evidence that a single mutation is driving some of the key adverse effects associated with ZIKV infection. Phylogenetic evidence suggests that the S139M prM mutation arose sometime after the virus had reached Asia but before it arrived in French Polynesia (Cao-Lormeau *et al.*, 2014).

The flavivirus NS1 protein has multiple functions, including assisting the virus with RNA replication, evading the host immune system, and causing vascular permeability in its secreted form (Glasner *et al.*, 2018). The mutation A188V in ZIKV NS1 conferred two different abilities on the virus. First, the mutation was shown to interfere with cell signalling that is part of the interferon pathway (Xia *et al.*, 2018). When a virus infects a human, one of the critical, front-line responses is the interferon pathway, which activates a variety of proteins to target and destroy the invading virus. In the tug-of-war between the virus and the human immune system, the NS1 A188V mutation helps the virus push back against these interferon proteins and survive to inflict damage on the human host. Second, the NS1 A188V mutation has also been shown to enhance the infectivity of ZIKV in *Aedes aegypti* mosquitoes (Liu *et al.*, 2017). The A188V mutation is thought to have arisen in Southeast Asia in the early 2000s (Delatorre *et al.*, 2017).

By two years after the start of the Zika epidemic in the Americas, researchers had identified the causative virus and functional mutations that were critical to its spread across the continents. However, the emergence and severity of Zika disease is explained by more than just mutations in the virus's genome. Other biological factors thought to be involved in the disease process include individuals' prior history of DENV infection and their genetic background.

Beyond biology, a variety of social, political, and environmental factors have also played a part in driving the patterns of Zika disease emergence. For example, co-morbidities, the other diseases that people might already have when they get infected with ZIKV, are thought to exacerbate the symptoms of Zika. Access to contraception, abortion services, and family planning options also have a role to play. Across the Americas, the legality of these services varies widely, and country-specific norms may be as important to accessing these options as laws and regulations. For political reasons, some countries are not reporting, or are downplaying, the number of microcephaly cases that occurred. This systematic under-reporting makes it difficult to characterise the true impact and burden that Zika has had on the health of the continent. Finally, it is also important to consider local environmental factors, which will be discussed in the next section.

To explore just one of these factors, it is illustrative to consider the impact of 'upward migration', or social mobility, on Zika disease. A recent study was carried out in northeastern Brazil, the epicentre of ZIKV-associated microcephaly cases (Vieira de Souza *et al.*, 2018). Researchers compared the rates of microcephaly in different neighbourhoods defined by their socioeconomic levels. While the richest neighbourhoods had an average rate of 12 microcephalic births per 10,000 newborns, the rate was 67 per 10,000 newborns in the poorest neighbourhoods. This difference was statistically significant, meaning that it was very unlikely to be due to chance alone. A similar spatial relationship between the occurrence of microcephaly and poor areas may be present in Nicaragua as well, and our research group is in the process of examining this hypothesis. For now, the underlying reason for this poverty–microcephaly association remains unknown. Lack of abortion services, poor nutritional status, and a variety of other factors could underlie this phenomenon, but further research is needed to understand precisely why microcephaly is associated with poverty.

Migration around Cities: Environmental Hot Spots

This section explores how infections and instances of disease are spread throughout a population and how that dissemination is measured. Before discussing the spatial migration of DENV and ZIKV at the level of a city, it is important to understand the design of our Nicaraguan studies and how our research team is able to collect so much data.

Most of my laboratory's immunology and epidemiology research stems from a large paediatric cohort study in Managua, Nicaragua. Starting in

2004, we began enrolling children two years of age and older every year, and we follow them until they turn 15 years of age (Kuan *et al.*, 2009). As part of the study, we provide medical care to 3,700 children year-round. The children are examined at our health centre for a variety of health conditions, but we are particularly interested in febrile illnesses because DENV, ZIKV, and the other viruses we study cause fever. When we suspect a dengue or Zika case, we run laboratory tests to confirm it. In addition, every year, participating children donate one healthy blood sample, which is then tested for antibodies against DENV, ZIKV, and other viruses. From these blood samples, we are able to determine which viruses the children were infected by during the previous year. When all of the blood samples from each year are analysed, we can reconstruct a child's infection history and assess whether any factors increased or decreased the risk of becoming infected. Using the same analytical framework, it is possible to identify risk factors for disease occurrence and symptom severity.

Because our study is so large, and because we are very careful to ensure the accuracy of our data, the results from our study are a good reflection of what happens in Managua. In particular, because we follow our participants with such intensity, we can use our study to learn how widespread an epidemic might have been. For example, after analysing data from 2004 to 2017, we know that a typical dengue outbreak in our study population results in about 15 cases of dengue every two weeks during the epidemic season, although this varies year-to-year (Figure 5.2). We can use this number as a reference point for other viral outbreaks. When chikungunya virus, which is also spread by the same *Aedes* mosquitoes, entered into our study site, it caused 19 cases of disease at the peak of its first epidemic wave in 2014 and 68 cases of disease at the peak of its second wave in 2015. These two epidemics, especially the one in 2015, were huge! We had never seen an epidemic of that ferocity in our study site before. When Zika arrived in 2016, it was even worse! At its peak, 83 cases came to the health centre every two weeks. It was an incredibly powerful epidemic that, for 2.5 months, completely overwhelmed the city's public health system.

As part of our research protocols, we record the GPS location (longitude and latitude coordinates) of each participant's home. This allows our research team to map the location of each infection as well as each instance of disease. For symptomatic infections that are reported to the health centre, we also have a date of initial symptom onset. This allows us to plot the time course of symptomatic infections and visualise where and when each case was located in both time and space.

Of the four DENV serotypes, DENV-1, -2, and -3 are present in Nicaragua. In any given year, only one serotype tends to dominate. A time lapse of disease data shows a scattering of dengue cases in the first half of the year, followed by a clear seasonal trend: dengue cases are concentrated in the September–December time period, which coincides with the rainy season when mosquitoes breed the most. After the rainy season is over, dengue cases drop dramatically. Over the years, the dominant serotype changes. For example, when DENV-2 had dominated for a few years, enough individuals had been exposed to it to generate population-wide immunity. This means the likelihood of DENV-2 infections fell, and a serotype to which a substantial fraction of the population was susceptible became dominant; in this case, DENV-3. The DENV-3 epidemic lasted a few years and was then replaced by DENV-1. This cycle repeats over and over again.

In contrast to the relatively slow-moving dengue epidemics, the Zika epidemic of 2016 was fast and impacted almost every corner of our study site. At the time of occurrence of the epidemic, its full extent was unclear. We knew how many cases had been reported to the health centre, but we also knew that many were not reported – because Zika disease is relatively mild compared with dengue. Some participants may not have felt sick enough to go to the health centre where we would have recorded their illness. At the time, an important and unaddressed question revolved around the proportion of the population that had been infected by ZIKV. The whole region, and in particular the Ministry of Health, was interested in estimating this percentage because it would determine the percentage of the population that would remain susceptible to a future epidemic. If a significant portion of the population lacked immunity, the region would likely experience another epidemic soon, and this possibility worried many public health professionals at the time.

However, when the epidemic occurred, reliable tools to answer these questions did not exist. All of the laboratory tests we had previously used to measure infection history, and hence immune status, employed blood-based antibody detection. Because of cross-reactivity, these tests often failed to distinguish between antibodies to DENV and antibodies to ZIKV; therefore, it was difficult to tell who had been infected by which virus.

To develop a new test that could overcome antibody cross-reactivity, we started with very specific antibodies to the ZIKV NS1 protein that we obtained from collaborators in Switzerland. The key idea behind this new test, called the Zika NS1 blockade-of-binding (BOB) enzyme-linked immunosorbent assay (ELISA), is quite simple: the test examines how strongly antibodies in someone's blood bind to a specific site on the Zika NS1 protein. During a ZIKV infection, antibodies are made to the NS1 protein that ZIKV produces. However, because these antibodies are made specifically to counter the ZIKV NS1 protein, they will bind more strongly and more specifically to the protein than will antibodies that were made to the DENV NS1 protein during a previous DENV infection. This difference in binding strength and site allows us to distinguish the presence of antibodies against DENV NS1 and antibodies against ZIKV NS1 in someone's blood. After validating the test in six countries (including Nicaragua), we found that the test could correctly identify 95% of those with a ZIKV infection as being ZIKV-positive, and could classify 96% of those without a ZIKV infection as being ZIKV-negative, starting three weeks after infection (Balmaseda *et al.*, 2017, 2018).

After we had shown that the test worked as expected, we used it on all of the 3,740 samples in our paediatric cohort study, as well as a separate study of adults, to measure how many participants had been infected with ZIKV during the single epidemic (Zambrana *et al.*, 2018). We found that 36% of the children and 56% of the adults had been infected in a three-month epidemic period – which is quite impressive! Overall, 46% of participants were ZIKV-positive. This again demonstrated that the epidemic had been explosive, but the silver lining is that such a widespread epidemic in a completely susceptible population generated widespread immunity to the virus. Such a high level of immunity likely precludes the possibility of another Zika epidemic occurring in Managua and likely throughout much of Latin America. And, in fact, considerably fewer cases have been reported in the aftermath of the Zika epidemic, again owing to the high percentage of immunity in the population.

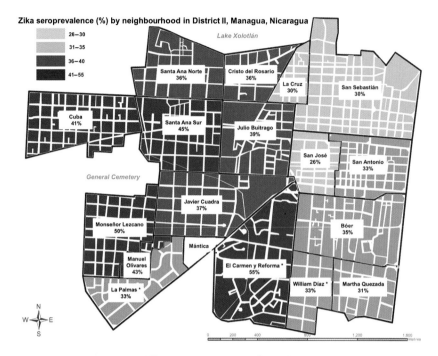

FIGURE 5.3 Zika seroprevalence in the Pediatric Dengue Cohort Study neighbourhoods in Managua, Nicaragua, after the 2016 Zika epidemic. Figure by Jairo Carey.

Our team then combined the infection data from the Zika NS1 BOB ELISA and the GPS points of the study participants' homes to map the infections spatially. The analysis showed a curious spatial pattern: infections tended to occur in the western half of our study site, and the highest prevalence estimates were in those western neighbourhoods that surrounded a cemetery (Figure 5.3) (Zambrana *et al.*, 2018). It was only after conducting these analyses that we realised that the cemetery was likely acting as an environmental source of infections. Whenever residents in the area brought flowers in vases with clean water into the cemetery, they were accidentally making that vase a potential breeding ground for *Aedes* mosquitoes, and they were accidentally providing the female mosquitoes with a delicious blood meal!

Cemeteries are important locales for mosquito breeding and development because the mosquito life cycle is facilitated in these settings. After a female *Aedes* mosquito has deposited eggs into clean water, it takes about three days for the larvae to hatch, then three or four days for the larvae to become pupae, and two more days for the pupae to develop into adult mosquitoes (Centers for Disease Control and Prevention, 2012). Cemetery groundskeepers do not wash out each and every vase in a vast cemetery, let alone dry each little puddle of water that accumulates during the rainy season, when it rains regularly. Consequently, the entire cemetery turns into a perfect habitat for mosquitoes, which infect individuals living around the cemetery as well as people and animals visiting the cemetery when the mosquitoes seek a blood meal.

In contrast, mosquitoes' life cycle is often interrupted in the home. Although people may store clean water in barrels, that water is being used every day for washing the dishes, cleaning the patio, showering, etc. The constant emptying and refilling of these barrels interrupts the mosquitoes' developmental cycle, which also breaks the transmission cycle of the disease. If eggs cannot become adult mosquitoes, adult mosquitoes cannot spread infections. Media campaigns and community education drives have taught people in Managua to interrupt dengue and Zika transmission in their homes. Empowering the local population with knowledge about the mosquito life cycle has taught them how they can directly prevent future cases of Zika and dengue (Arostegui *et al.*, 2017).

The spatial analysis also revealed a hotspot of infections in the interior of our study site, far enough away from the cemetery that infections were likely due to a different source. A site visit by members of our team discovered a tyre shop a few dozen feet from the centre of the high-risk area that sells new tyres and refurbishes old ones. A variety of other tyre shops and tyre-repair workshops are also located nearby. As explained previously, it takes only a little bit of water for a tyre to become a mosquito breeding ground. Given how often it rains in Managua, tyre shops can quickly become environmental sources of mosquito-borne infections if precautions are not taken. Our approach to solving this problem is evidence-based community-derived interventions – we inform community leaders of the problem, giving them knowledge about the mosquito life cycle, and they themselves come up with amazing solutions (Arostegui *et al.*, 2017). Most often, residents took it upon themselves to fill old tyres with earth; these tyres were then used to make staircases on the side of a muddy hill or to make planters for a garden. This approach once again empowers the local community and gives them agency to solve the problem in a community-appropriate manner.

In decades of thinking about how mosquito populations can be controlled, the World Health Organization, the Centers for Disease Control and Prevention, and non-governmental organisations generally did not think to *use* the old tyres. Prior recommendations had focused on disposing of the tyres in centralised locations (where they can continue to act as breeding sources for surrounding neighbourhoods) and burning the tyres in industrial facilities (Blackman and Palma, 2002). However, when outsiders came into communities and told the locals that their tyres were junk that needed to be burned, community members were rightfully offended. This kind of top-down messaging was counter-productive and did not engender trust between the affected population and the public health workers, even though the latter were well-intentioned. However, evidence-based community-derived interventions bring in members of the local population as true partners to jointly solve long-standing problems (Ledogar *et al.*, 2017). By working hand-in-hand with community members and leaders, by listening to their concerns and ideas, we have been able to make much progress in reducing the burden of mosquito-borne infections (Andersson *et al.*, 2015). In future, this model of community involvement will only become a greater and more powerful tool to improve the daily lives of the Managua community and beyond.

Migration within the Human Host: Fitness, Replication, and Immune Evasion

This final section will concentrate on the migration, or movement, of viruses throughout the human body and how different factors contribute to viral invasion and disease onset. As in previous sections, it will also illustrate the migration of mutations over time, or evolution, but on a much smaller timescale and spatial scale – evolution during acute DENV and ZIKV infections within a single infected individual.

When the scientific community realised what we were up against with an emerging virus that unexpectedly caused congenital defects, researchers with distinct but relevant areas of expertise rapidly began to collaborate across disciplines to address critical questions about Zika. For example, we combined our expertise in flavivirus research with that of Dr Lenore Pereira, a colleague at the University of California, San Francisco, who is an expert on the transmission of cytomegalovirus during pregnancy into the developing fetus. Together, we investigated how ZIKV is able to infect different types of placental cells and cross the human placental barrier. Although the placenta is a protective organ that isolates the fetus from most infections, somehow ZIKV is able to penetrate it, and only by using primary human cells and placental explants (derived from miscarriages that would otherwise be discarded at a hospital) have scientists begun to understand how and why this occurs.

Some of the work that we have done focuses on the chorionic membrane, which surrounds the placenta, and the chorionic villi, which anchor the placenta to the uterine wall and provide nutrients to the fetus. Using a technique called immunofluorescence microscopy, it is possible to identify different kinds of cells in the placental explant that ZIKV infects. For example, ZIKV NS3 protein is found in certain cell types, indicating that the virus can infect those cell types. The NS3 protein is made only when the virus is replicating, so detecting it via immunofluorescence microscopy tells us where the virus has established an active infection.

Other work from our groups demonstrated the importance of Hofbauer cells to ZIKV migration through the placenta. Hofbauer cells are fetal macrophages, white blood cells that protect the fetus against pathogens such as bacteria and viruses. Macrophages are regularly infected by DENV and ZIKV, and we have found that, when ZIKV infects these cells in the placenta, the infected Hofbauer cell is able to help the virus migrate through the placental tissue and produce more progeny virions (Tabata et al., 2016). We also found differences in the ability of ZIKV strains to infect placental cells during placental development (Tabata et al., 2018).

Importantly, we characterised the routes through which we believe ZIKV is able to cross into the placenta. These experiments used explants and cells that were derived from different parts of the placental structure. At the time, researchers believed that the virus was entering the placenta only through the chorionic villi. However, our research uncovered a potential second route through the amniotic membrane, the membrane that surrounds the developing fetus and contains the amniotic fluid

(Tabata *et al.*, 2016). We then identified drugs that could block the migration of ZIKV into the placenta. Our experiments provided a proof of concept, as we used drugs approved for animal use, which is different from using drugs made for human use in pregnant women. However, having identified the cellular receptors potentially responsible for the migration of ZIKV into the human placenta, our research provides a framework for future development of drugs that block these specific routes and thereby prevent ZIKV infection of the fetus.

Up to this point, I have described mutations that uniformly impact entire strains of viruses all at once; however, the reality is more complicated. When an individual virus infects a cell, it will make millions of copies of itself, these copies will then exit the cell, and each one will attempt to infect other cells. Owing to the high mutation rate of RNA viruses, each individual copy will vary from any other copy by at least one mutation. Therefore, a cloud of very similar viruses with slight differences in their genetic makeup exist within one particular infected person (called 'intra-host diversity' or 'quasispecies') (Domingo *et al.*, 2006).

Before our studies on this topic, intra-host genetic diversity was thought to be relevant only for viruses, such as HIV and hepatitis C virus, that persist in the body for many years. It was believed that only such a long timeframe could allow many different mutations to arise. A virus like DENV or ZIKV, which circulates in the body for only a week or so before the immune system eliminates it, was not thought to give rise to much intra-host diversity in the human host.

We set out to examine whether the common wisdom regarding DENV and intra-host genetic diversity was true (Parameswaran *et al.*, 2012, 2017). We started with venous blood samples that had been collected from hospitalised Nicaraguan children who were suffering from dengue. We analysed the viruses within two components of the blood sample: the circulating white blood cells (peripheral blood mononuclear cells) and the serum, the fluid part of blood that does not contains any cells or clotting factors. Using intra-host sequencing techniques, we examined the RNA corresponding to the entire genome of every individual copy of DENV-3 and compared the genetic sequences in order to understand even minor differences among them.

When we compared mutations in the envelope (E) protein occurring in DENV-3 from serum with E protein mutations occurring in DENV-3 from circulating white blood cells, we found, to our surprise, that there was very little overlap between the two. In fact, we observed mutation hotspots for DENV-3 from the bloodstream that were different from the mutation hotspots in viruses found in circulating white blood cells. This suggested that the viral mutations inside the circulating white blood cells are very different from those outside the cells, which implied that the viruses inside the cells are not the ones that are populating the bloodstream. The dengue field had become so used to working only with venous blood that scientists assumed that, since DENV infects white blood cells, the virus in the bloodstream was derived from these infected cells. However, our research showed that the virus in the bloodstream was likely generated from other white blood cell populations in the spleen or other organ systems. Only by looking at the genetics of the viral quasispecies were we able to deduce the existence of more than one source of viruses in the human body during a DENV infection.

Moreover, we found that the number of mutations in the E protein from a first DENV infection was significantly higher than the number of E protein mutations during a second DENV infection. We discovered that the strong immune response elicited during a second infection exerted micro-evolutionary pressure on the viruses, reducing their mutation rate. Putting all the evidence together, we had a strong case regarding the occurrence of mutations within a very short time interval, and, interestingly, some evidence that the human immune system was partially driving these micro-evolutionary processes.

Our research indicated that the RNA sequence of the E protein, as well as the prM protein (also present on the virion surface), from different patients had the same hotspot for mutations. That 80% of our patients had the same mutation in the same location during the course of infection was surprising and constituted clear evidence of convergent evolution. When the function of the hotspot undergoing convergent evolution was analysed, we found that all the viruses developed the same mutation in the E protein – which resulted in a diminished capacity to replicate, meaning that the mutation reduced their overall fitness. However, this mutation kept arising over and over again in the same hotspot because, as it turned out, that particular mutation allowed the viruses to escape the antibody response. In the battle between the virus and the host's immune system, mutations that allowed the virus to survive in the presence of a strong immune response were being selected for in different people, despite incurring a reduction in viral fitness. However, because of the reduced replicative capacity, the viruses with this mutation were never able to reach 'consensus' level and take over the viral population – thus, the only way to observe this microscopic battle was by looking at the intra-host diversity of viral sequences. This form of migration of muta- tions during acute viral infection had not previously been observed in such a short amount of time, revealing just how much we have yet to learn about dengue disease progression and evolution in the human host.

Most dengue studies are based on samples of venous blood because DENV persists there for around five days. However, ZIKV tends to disappear from blood within three or four days, so clinicians and researchers may miss detecting an infection if they rely on blood-based tests. It turns out that ZIKV RNA can be detected in urine for longer periods of time, up to several weeks in some cases (Burger-Calderon *et al.*, 2018). We conducted a different study, called an index cluster study, to characterise the amount and persistence of ZIKV in different bodily fluids (Burger-Calderon *et al.*, 2018). In our study, we identified 33 ZIKV- infected children and then enrolled 109 members of their households. We returned to the homes to collected samples at day 1, day 3–4, day 6–7, day 9–10, and day 21. By examining these bodily fluids longitudin- ally, we found that ZIKV RNA levels in blood peaked two days after Zika disease onset but fell below detection levels two days thereafter. ZIKV RNA persisted in saliva for a week and was detectable in urine up to three weeks after the onset of symptoms. This knowledge directly informs future clinical diagnostics, as the doctors we work with now know that, among the easily obtainable sample types, urine and saliva can be used as diagnostic specimens as well.

Finally, the persistence of ZIKV in different bodily fluids can be analysed in terms of intra-host genetic diversity. Such analyses will reveal the migration of mutations across different fluids sampled on the same day in the same person as well as within the same bodily fluid across time. This work is still ongoing, but, on the basis of prior work with

DENV, we expect to find different mutations arising in the different virus populations that infect different bodily compartments. We also expect to find some degree of convergent evolution within the different virus populations, but further work is needed to test these hypotheses explicitly.

Conclusion

In the face of a rapidly emerging Zika pandemic that was explosive not only in terms of numbers but also in terms of biological and clinical conundrums, we, and the scientific community more broadly, attempted to respond immediately by answering as many questions as we could across the board. Here we were helped by our 20-year history of studying dengue, as well as by the holistic approach to scientific research that our programme embodies, which addresses multiple disciplines across the spectrum (Figure 5.4) in a collaborative fashion, combining research at UC Berkeley and in Nicaragua. Thus, we were able to tackle a number of what we call the 'zillion Zika questions' simultaneously and essentially pivot rapidly and port our expertise in dengue laterally to address many

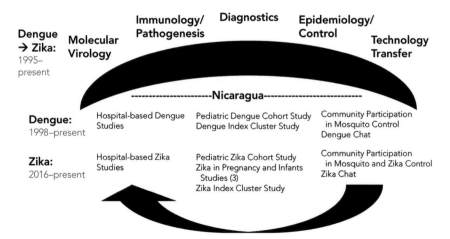

FIGURE 5.4 Integrated dengue–Zika research programme. Figure by Eva Harris and Fausto Bustos Carrillo.

disparate, urgent issues in Zika, ranging from pathogenesis to diagnostics to epidemiology to community-based participatory research.

The results of all that labour have helped us understand the migration of Zika and the causative virus across continents, around cities, and within the human host. Genetic studies have helped the scientific community uncover the specific mutations that arose to confer the ZIKV pandemic strain with different functions regarding immune evasion, infectivity in the mosquito, and disruption of neuronal development, among other things. From an evolutionary perspective, this has helped us understand how natural selection operates on the ZIKV genome to make it a more adaptive virus.

At the level of the city, in contrast to the typically slower patterns of migration of dengue epidemics through our cohort, Zika moved incredibly fast through our population and infected about half the population in less than three months. However, this left about half of the overall population with immunity, which will likely preclude a similarly large epidemic in the near future. Having developed new laboratory methods to identify ZIKV infections even in a dengue-endemic area, we identified spatial risk factors and used that information to develop possible strategies to mitigate future instances of infection. For instance, we now know that the cemetery next to our study site is a source for a large number of mosquitoes and thus infections, so we are working with members of the community to implement sustainable solutions that work.

Research tracking the viral invasion of the human host showed how ZIKV physically migrates into the placenta and which routes the virus could use to infect the developing fetus. Research into intra-host genetic diversity has also shown genetic migration of DENV and ZIKV RNA genomes operating on a very short timescale. Micro-evolution and natural selection were shown to be occurring at the interface of viral fitness and the selective pressures exerted by the human immune response.

To a great extent, science can be conceptualised as a kind of continual migration of evidence-based theories and ideas to explain phenomena in the world. Observing phenomena, forming hypotheses, collecting data, and analysing data to test competing hypotheses are critical to the scientific method. However, integrating results across disciplines to arrive at universal truths is just as critical to achieving the ends of the scientific method. Ultimately, societal improvement relies on the migration of ideas across time and space in the scientific realm in order to move forwards. A quotation (possibly incorrectly attributed to Darwin!) sums up this notion quite nicely: 'In the long history of humankind (and animal kind, too) those who learned to collaborate and improvise most effectively have prevailed.' To answer the most pressing public health questions effectively, scientific collaboration is critical on so many different levels that the scientific endeavour often, and rightly so, becomes a collaboration of disparate disciplines joining together synergistically.

However, the current constant drive to publish paper after paper can lead scientists into intellectual silos, where collaboration across disciplines and big-picture thinking are sacrificed for the immediacy of finishing the next project. I believe that this sort of shortsighted incentive structure and intellectual framework represents a dangerous approach to the way science is carried out.

My particular research programme provides a counterexample to the 'siloed' approach to science. As mentioned above, I strongly believe that to approach a disease and truly understand it, to interrogate it in its entirety, a holistic, integrated approach is necessary (Figure 5.4). Nothing short of bringing the analytical capacity of multiple disciplines to bear on the same question will lead to a full understanding of disease processes and effective control strategies.

Science as a collaborative effort is incredibly exciting in many ways. Nicaragua, which has been a huge part of my life, has also imbued it with so much meaning. Advancing the collective scientific understanding and making substantial improvements in public health has been a rewarding and fascinating experience. Additionally, part of my work in Nicaragua over the last 30 years has been helping to build the scientific infrastructure and train scientists in-country as well as in my laboratory at UC Berkeley. Instead of parachuting scientists in to solve a health crisis, the existence of a robust scientific and public health infrastructure in Nicaragua allows the country to handle its own problems locally. Taking local solutions and reporting those approaches to the wider international community advances science and helps communities around the world adopt effective disease control strategies.

This chapter and the underlying studies would not have been possible without the support of the wonderful members of my laboratory at the University of California, Berkeley, both in our Pathogenesis Group and in our Immunology Group. Further, approximately 200 incredibly talented, dedicated, and hard-working people work on our projects in Nicaragua. These include personnel working at the Laboratorio Nacional de Virología; our health centre, the Centro de Salud Sócrates Flores Vivas; and our hospital, Hospital Infantil Manuel de Jesús Rivera, 'La Mascota', all of which form part of the Ministry of Health. Additionally, the work of the Sustainable Sciences Institute, our non-profit organisation based both in San Francisco and in Nicaragua, has been invaluable to our ongoing research projects and to our overall vision. Our studies receive much support from the Nicaraguan Ministry of Health as well as from members of my laboratory and colleagues, in particular Dr Josefina Coloma and Dr Aubree Gordon, lifelong collaborators also committed to using science to make the world a better place. We are grateful to Dr Leah Katzelnick, Jairo Carey, and Paolo Harris Paz for contributions to the figures in this chapter. And, of course, none of this research would be possible without the participating children and their families, who give of their time and effort to contribute to science and help advance the cause of a better, healthier planet.

References

Al Awaidy, S. T., Al Obeidani, I., Bawikar, S., *et al.* (2014). 'Dengue epidemiological trend in Oman: a 13-year national surveillance and strategic proposition of imported cases'. *Trop. Doct.*, 44(4), 190–5. https://doi.org/10.1177/0049475514541650

Andersson, N., Nava-Aguilera, E., Arosteguí, J., *et al.* (2015). 'Evidence based community mobilization for dengue prevention in Nicaragua and Mexico (Camino Verde, the Green Way): cluster randomized controlled trial'. *BMJ*, 351, h3267. https://doi.org/10.1136/BMJ.H3267

Arosteguí, J., Ledogar, R. J., Coloma, J., *et al.* (2017). 'The Camino Verde intervention in Nicaragua, 2004–2012'. *BMC Public Health*, 17(S1), 406. https://doi.org/10.1186/s12889-017-4299-3

Aubry, M., Teissier, A., Huart, M., *et al.* (2017). 'Zika virus seroprevalence, French Polynesia, 2014–2015'. *Emerg. Infect. Dis.*, 23(4), 669–72. https://doi.org/10.3201/eid2304.161549

Balmaseda, A., Stettler, K., Medialdea-Carrera, R., *et al.* (2017). 'Antibody-based assay discriminates Zika virus infection from other flaviviruses'. *Proc. Natl Acad. Sci. U.S.A.*, 114(31), 8384–9. https://doi.org/10.1073/pnas.1704984114

Balmaseda, A., Zambrana, J. V., Collado, D., *et al.* (2018). 'Comparison of four serological methods and two reverse transcription–PCR assays for diagnosis and surveillance of Zika virus infection'. *J. Clin. Microbiol.*, 56(3), e01785-17. https://doi.org/10.1128/JCM.01785-17

Barba-Spaeth, G., Dejnirattisai, W., Rouvinski, A., *et al.* (2016). 'Structural basis of potent Zika–dengue virus antibody cross-neutralization'. *Nature*, 536(7614), 48–53. https://doi.org/10.1038/nature18938

Beatty, P. R., Puerta-Guardo, H., Killingbeck, S. S., *et al.* (2015). 'Dengue virus NS1 triggers endothelial permeability and vascular leak that is prevented by NS1 vaccination'. *Sci. Transl. Med.*, 7(304), 304ra141. https://doi.org/10.1126/scitranslmed.aaa3787

Bhatt, S., Gething, P. W., Brady, O. J., *et al.* (2013). 'The global distribution and burden of Dengue'. *Nature*, 496(7446), 504–7.

Blackman, A., and Palma, A. (2002). 'Scrap tires in Ciudad Juárez and El Paso: ranking the risks', https://ideas.repec.org/p/ags/rffdps/10583.html (accessed 9 October 2019).

Burger-Calderon, R., Gonzalez, K., Ojeda, S., *et al.* (2018). 'Zika virus infection in Nicaraguan households'. *PLoS Negl. Trop. Dis.*, 12(5), e0006518. https://doi.org/10.1371/journal.pntd.0006518

Cao-Lormeau, V.-M., Roche, C., Teissier, A., *et al.* (2014). 'Zika virus, French Polynesia, South Pacific, 2013'. *Emerg. Infect. Dis.*, 20(6), 1084–6. https://doi.org/10.3201/eid2006.140138

Centers for Disease Control and Prevention. (2012). 'Life cycle: the mosquito', www.cdc.gov/dengue/resources/factSheets/MosquitoLifecycleFINAL.pdf (accessed 9 October 2019).

(2018). 'Congenital Zika syndrome & other birth defects', www.cdc.gov/pregnancy/zika/testing-follow-up/zika-syndrome-birth-defects.html (accessed 3 April 2018).

Darwin, C. (1859). *On the Origin of Species by Means of Natural Selection, or the Preservation of Favoured Races in the Struggle for Life*. London: John Murray.

Delatorre, E., Mir, D., and Bello, G. (2017). 'Tracing the origin of the NS1 A188V substitution responsible for recent enhancement of Zika virus Asian genotype infectivity'. *Mem. Inst. Oswaldo Cruz*, 112(11), 793–5. https://doi.org/10.1590/0074-02760170299

Dick, G. W. A., Kitchen, S. F., and Haddow, A. J. (1952). 'Zika virus. I. Isolations and serological specificity'. *Trans. R. Soc. Trop. Med. Hyg.*, 46(5), 509–20.

Domingo, E., Martin, V., Perales, C., *et al.* (2006). 'Viruses as quasispecies: biological implications'. *Curr. Top. Microbiol. Immunol.*, 299, 51–82. www.ncbi.nlm.nih.gov/pubmed/16568896

Duffy, M. R., Chen, T.-H., Hancock, W. T., *et al.* (2009). 'Zika virus outbreak on Yap Island, Federated States of Micronesia'. *N. Engl. J. Med.*, 360(24), 2536–43. https://doi.org/10.1056/NEJMoa0805715

Ehrenkranz, N. J., Ventura, A. K., Cuadrado, R. R., Pond, W. L., and Porter, J. E. (1971). 'Pandemic dengue in Caribbean countries and the Southern United States – past, present and potential problems'. *N. Engl. J. Med.*, 285(26), 1460–9. https://doi.org/10.1056/NEJM197112232852606

Fagbami, A. H. (1979). 'Zika virus infections in Nigeria: virological and seroepidemiological investigations in Oyo State'. *J. Hyg. (Lond.)*, 83(2), 213–219. www.ncbi.nlm.nih.gov/pubmed/489960

Faria, N. R., Quick, J., Claro, I. M., *et al.* (2017). 'Establishment and cryptic transmission of Zika virus in Brazil and the Americas'. *Nature*, 546(7658), 406–10. https://doi.org/10.1038/nature22401

Glasner, D. R., Puerta-Guardo, H., Beatty, P. R., and Harris, E. (2018). 'The good, the bad, and the shocking: the multiple roles of dengue virus nonstructural protein 1 in protection and pathogenesis'. *Annu. Rev. Virol.*, 5(1), 101416-041848. https://doi.org/10.1146/annurev-virology-101416-041848

Guzman, M. G., Gubler, D. J., Izquierdo, A., Martinez, E., and Halstead, S. B. (2016). 'Dengue infection'. *Nat. Rev. Dis. Prim.*, 2, 16055. https://doi.org/10.1038/nrdp.2016.55

Guzman, M. G., Halstead, S. B., Artsob, H., *et al.* (2010). 'Dengue: a continuing global threat'. *Nat. Rev. Microbiol.*, 8(12), S7–S16. https://doi.org/10.1038/nrmicro2460

Guzmán, M. G., and Kourí, G. (2002). 'Dengue: an update'. *Lancet. Infect. Dis.*, 2(1), 33–42. www.ncbi.nlm.nih.gov/pubmed/11892494

Halstead, S. B., Nimmannitya, S., Yamarat, C., and Russell, P. K. (1967). 'Hemorrhagic fever in Thailand; recent knowledge regarding etiology'. *Jpn J. Med. Sci. Biol.*, 20 Suppl., 96–103. www.ncbi.nlm.nih.gov/pubmed/5301574

Hawley, W. A., Reiter, P., Copeland, R. S., Pumpuni, C. B., and Craig, G. B. (1987). '*Aedes albopictus* in North America: probable introduction in used tires from northern Asia'. *Science*, 236(4805), 1114–16. www.ncbi.nlm.nih.gov/pubmed/3576225

Kambhampati, S., Black, W. C., and Rai, K. S. (1991). 'Geographic origin of the US and Brazilian *Aedes albopictus* inferred from allozyme analysis'. *Heredity (Edinb.)*, 67(1), 85–94. https://doi.org/10.1038/hdy.1991.67

Kouri, G. P., Guzmán, M. G., and Bravo, J. R. (1987). 'Why dengue haemorrhagic fever in Cuba? An integral analysis'. *Trans. R. Soc. Trop. Med. Hyg.*, 81(5), 821–3. www.ncbi.nlm.nih.gov/pubmed/3450005

Kraemer, M. U. G., Sinka, M. E., Duda, K. A., *et al.* (2015). 'The global distribution of the arbovirus vectors *Aedes aegypti* and *Ae. albopictus*'. *eLife*, 4, e08347. https://doi.org/10.7554/eLife.08347

Kuan, G., Gordon, A., Avilés, W., *et al.* (2009). 'The Nicaraguan Pediatric Dengue Cohort Study: study design, methods, use of information technology, and extension to other infectious diseases'. *Am. J. Epidemiol.*, 170(1), 120–9. https://doi.org/10.1093/aje/kwp092

Lai, C.-Y., Tsai, W.-Y., Lin, S.-R., *et al.* (2008). 'Antibodies to envelope glycoprotein of dengue virus during the natural course of infection are predominantly cross-reactive and recognize epitopes containing highly conserved residues at the fusion loop of domain II'. *J. Virol.*, 82(13), 6631–43. https://doi.org/10.1128/JVI.00316-08

Lednicky, J., Madsen Beau De Rochars, V., El Badry, M., *et al.* (2016). 'Zika virus outbreak in Haiti in 2014: molecular and clinical data'. *PLoS Negl. Trop. Dis.*, 10(4), e0004687. https://doi.org/10.1371/journal.pntd.0004687

Ledogar, R. J., Arosteguí, J., Hernández-Alvarez, C., *et al.* (2017). 'Mobilising communities for *Aedes aegypti* control: the SEPA approach'. *BMC Public Health*, 17(Suppl. 1), 403. https://doi.org/10.1186/s12889-017-4298-4

Liu, Y., Liu, J., Du, S., *et al.* (2017). 'Evolutionary enhancement of Zika virus infectivity in *Aedes aegypti* mosquitoes'. *Nature*, 545(7655), 482–6. https://doi.org/10.1038/nature22365

McBride, C. S., Baier, F., Omondi, A. B., *et al.* (2014). 'Evolution of mosquito preference for humans linked to an odorant receptor'. *Nature*, 515(7526), 222–7. https://doi.org/10.1038/nature13964

Metsky, H. C., Matranga, C. B., Wohl, S., *et al.* (2017). 'Zika virus evolution and spread in the Americas'. *Nature*, 546(7658), 411–15. https://doi.org/10.1038/nature22402

Modhiran, N., Watterson, D., Muller, D. A., *et al.* (2015). 'Dengue virus NS1 protein activates cells via Toll-like receptor 4 and disrupts endothelial cell monolayer integrity'. *Sci. Transl. Med.*, 7(304), 304ra142. https://doi.org/10.1126/scitranslmed.aaa3863

Murray, N. E. A., Quam, M. B., and Wilder-Smith, A. (2013). 'Epidemiology of dengue: past, present and future prospects'. *Clin. Epidemiol.*, 5, 299–309. https://doi.org/10.2147/CLEP.S34440

Pan American Health Organization. (1971). 'Guide for the reports on the *Aedes Aegypti* eradication campaign in the Americas', http://iris.paho.org/xmlui/handle/123456789/18650 (accessed 9 October 2019).

Parameswaran, P., Charlebois, P., Tellez, Y., *et al.* (2012). 'Genome-wide patterns of intrahuman dengue virus diversity reveal associations with viral phylogenetic clade and interhost diversity'. *J. Virol.*, 86(16), 8546–58. https://doi.org/10.1128/JVI.00736-12

Parameswaran, P., Wang, C., Trivedi, S. B., *et al.* (2017). 'Intrahost selection pressures drive rapid dengue virus microevolution in acute human infections'. *Cell Host Microbe*, 22(3), 400–10. https://doi.org/10.1016/j.chom.2017.08.003

Pettersson, J. H.-O., Eldholm, V., Seligman, S. J., *et al.* (2016). How did Zika virus emerge in the Pacific islands and Latin America?' *mBio*, 7(5), 01239-16. https://doi.org/10.1128/mBio.01239-16

Pierson, T. C., and Diamond, M. S. (2012). 'Degrees of maturity: the complex structure and biology of flaviviruses'. *Curr. Opin. Virol.*, 2(2), 168–75. https://doi.org/10.1016/j.coviro.2012.02.011

Pompon, J., Morales-Vargas, R., Manuel, M., *et al.* (2017). 'A Zika virus from America is more efficiently transmitted than an Asian virus by *Aedes aegypti* mosquitoes from Asia'. *Sci. Rep.*, 7(1), 1215. https://doi.org/10.1038/s41598-017-01282-6

ProMED (2015). 'Doença misteriosa assusta população de Camaçari', www.promedmail.org/direct.php?id=20150325.3253769 (accessed 3 April 2018).

Puerta-Guardo, H., Glasner, D. R., and Harris, E. (2016). 'Dengue virus NS1 disrupts the endothelial glycocalyx, leading to hyperpermeability'. *PLOS Pathog.*, 12(7), e1005738. https://doi.org/10.1371/journal.ppat.1005738

Russell, T. L., Govella, N. J., Azizi, S., *et al.* (2011). 'Increased proportions of outdoor feeding among residual malaria vector populations following increased use of insecticide-treated nets in rural Tanzania'. *Malar. J.*, 10(1), 80. https://doi.org/10.1186/1475-2875-10-80

Saluzzo, J. F., Ivanoff, B., Languillat, G., and Georges, A. J. (1982). 'Enquête sérologique sur l'incidence des arbovirus parmi les populations humaines et simiennes du sud-est de la République Gabonaise [Serological survey for arbovirus antibodies in the human and simian populations of the south-east of Gabon]'. *Bull. Soc. Pathol. Exot. Filiales*, 75(3), 262–6. English abstract www.ncbi.nlm.nih.gov/pubmed/6809352

Sangkawibha, N., Rojanasuphot, S., Ahandrik, S., *et al.* (1984). 'Risk factors in dengue shock syndrome: a prospective epidemiologic study in Rayong, Thailand. I. The 1980 outbreak'. *Am. J. Epidemiol.*, 120(5), 653–69. www.ncbi.nlm.nih.gov/pubmed/6496446

Smithburn, K. C. (1952). 'Neutralizing antibodies against certain recently isolated viruses in the sera of human beings residing in East Africa'. *J. Immunol.*, 69(2), 223–34. www.ncbi.nlm.nih.gov/pubmed/14946416

Sprenger, D., and Wuithiranyagool, T. (1986). 'The discovery and distribution of *Aedes albopictus* in Harris County, Texas'. *J. Am. Mosq. Control Assoc.*, 2(2), 217–19.

Tabata, T., Petitt, M., Puerta-Guardo, H., *et al.* (2016). 'Zika virus targets different primary human placental cells, suggesting two routes for vertical transmission'. *Cell Host Microbe*, 20(2), 155–66. https://doi.org/10.1016/j.chom.2016.07.002

(2018). 'Zika virus replicates in proliferating cells in explants from first-trimester human placentas, potential sites for dissemination of infection'. *J. Infect. Dis.*, 217(8), 1202–13. https://doi.org/10.1093/infdis/jix552

Thein, S., Aung, M. M., Shwe, T. N., *et al.* (1997). 'Risk factors in dengue shock syndrome'. *Am. J. Trop. Med. Hyg.*, 56(5), 566–72. www.ncbi.nlm.nih.gov/pubmed/9180609

Thézé, J., Li, T., du Plessis, L., *et al.* (2018). 'Genomic epidemiology reconstructs the introduction and spread of Zika virus in Central America and Mexico'. *Cell Host Microbe*, 23(6), 855–64. https://doi.org/10.1016/j.chom.2018.04.017

Tognarelli, J., Ulloa, S., Villagra, E., *et al.* (2016). 'A report on the outbreak of Zika virus on Easter Island, South Pacific, 2014'. *Arch. Virol.*, 161(3), 665–8. https://doi.org/10.1007/s00705-015-2695-5

Vieira de Souza, W., de Fátima Pessoa Militão de Albuquerque, M., Vazquez, E., *et al.* (2018). 'Microcephaly epidemic related to the Zika virus and living conditions in Recife, northeast Brazil'. *BMC Public Health*, 18(1), 130. https://doi.org/10.1186/s12889-018-5039-z

Wilder-Smith, A., and Gubler, D. J. (2008). 'Geographic expansion of dengue: the impact of international travel'. *Med. Clin. North Am.*, 92(6), 1377–90. https://doi.org/10.1016/j.mcna.2008.07.002

Wilson, M., and Chen, L. (2002). 'Dengue in the Americas'. *Dengue Bull.*, 26, 44–61.

World Health Organization (1997). *Dengue Haemorrhagic Fever: Diagnosis, Treatment, Prevention, and Control*, 2nd edn. Geneva: World Health Organization, www.who.int/csr/resources/publications/dengue/Denguepublication/en/ (accessed 10 October 2019).

(2016a). 'WHO Director-General summarizes the outcome of the Emergency Committee regarding clusters of microcephaly and Guillain–Barré syndrome'. www.who.int/en/news-room/detail/01-02-2016-who-director-general-summarizes-the-outcome-of-the-emergency-committee-regarding-clusters-of-microcephaly-and-guillain-barré-syndrome (accessed 2 August 2018).

(2016b). 'Zika virus and complications: questions and answers', www.who.int/features/qa/zika/en/ (accessed 9 October 2019).

(2016c). 'Fifth meeting of the Emergency Committee under the International Health Regulations (2005) regarding microcephaly, other neurological disorders and Zika virus', www.who.int/en/news-room/detail/18-11-2016-fifth-meeting-of-the-emergency-committee-under-the-international-health-regulations-(2005)-regarding-microcephaly-other-neurological-disorders-and-zika-virus (accessed 2 August 2018).

Xia, H., Luo, H., Shan, C., *et al.* (2018). 'An evolutionary NS1 mutation enhances Zika virus evasion of host interferon induction'. *Nat. Commun.*, 9(1), 414. https://doi.org/10.1038/s41467-017-02816-2

Yuan, L., Huang, X.-Y., Liu, Z.-Y., *et al.* (2017). 'A single mutation in the prM protein of Zika virus contributes to fetal microcephaly'. *Science*, 358(6365), 933–6. https://doi.org/10.1126/science.aam7120

Zambrana, J. V., Bustos Carrillo, F., Collado, D., *et al.* (2018). 'Seroprevalence, risk factor, and spatial analysis of Zika virus infection after the 2016 epidemic in Managua, Nicaragua'. *Proc. Natl. Acad. Sci.*, 115(37), 9294–9. www.pnas.org/content/115/37/9294

6 The Partition of India and Migration

KAVITA PURI

We all have a story, a narrative. Friends and family members may know it, and it may get passed down the generations. Sometimes stories are all we have of a past; they may be dramatic – witnessing historic moments or disaster; they can be life-changing and are often ordinary. They can knit a sense of self, identity, and even bind a nation.

But sometimes these stories are too painful to talk of. The ones in this chapter are a very British story, yet one so many people have barely heard of. Dotted around homes in Britain are people who bore witness to one of the most violent upheavals of the twentieth century – they are former subjects of the British Raj who are now British citizens. They are complex stories from a generation who are vanishing. These people were affected by the end of empire, when the Indian subcontinent was divided into Hindu-majority India and Muslim-majority Pakistan 70 years ago. 'Partition' as it is known in Britain. It is one of the largest migrations – outside war and famine – in human history.

Millions who feared being a minority in their new country left their homes. Hindus and Sikhs to India, Muslims to Pakistan. I say it is a British story not just because of the 400-year-old connection between the Indian subcontinent and Britain, but because some of the people who had suffered and witnessed this violent birth of two nations migrated to post-war Britain – and their children and grandchildren are part of the fabric of contemporary British society. But the stories of that first generation who now live in Britain – of lands lost and friends never to be seen again – the *lived* experience of Partition; these are only just beginning to be told to public audiences in Britain.

I should declare a personal interest at this point. My obsession began because I am the daughter of a Partition survivor. My father was born into a Hindu family in Lahore in 1935 and spent the first 12 years of his life there. He travelled to Moga – in Eastern Punjab. He left Lahore with his mother and siblings in the months before Partition, when the communal violence in the city was getting too bad. All we knew from this time was that their Muslim neighbours in Lahore protected my family as much as they could – and that in Moga, once the area had been awarded to India, he witnessed terrible atrocities against the Muslim population there. These were the only scraps of information I could glean. But that was the extent of it. And over the 10 or so years I had been interviewing my father – mostly for our family archive, and partly for a number of Radio 4 programmes – I knew Partition was a definite no-go area. He would happily talk about his early life in Lahore, life in India post-1947, coming to Britain, and so on. But when we got to Partition he became monosyllabic, 'Why do you want to talk of that,' he would shrug.

I would have conversations with friends who were of British South Asian heritage where we would compare what we knew (which was not very much) from our parents of those Partition days. I knew there must be so many stories out there waiting to be excavated. I wanted to find them before it was too late. As part of a British Broadcasting Corporation (BBC) project called Partition Voices – whose main component was a Radio 4 series – I travelled across Britain interviewing people about their Partition memories. These interviews are being kept by the British Library Sound Archive. We contacted organisations and local papers across the country, academics, and faith leaders, and we did call-outs on the BBC Asian network and on Indian-language radio. Once we started looking, we realised the stories were all around us. We spoke to many, many people. We couldn't interview everyone – in the end we spoke to, at length, over 40 people on the record. We wanted, as far as possible, the full range of experience, of course Hindu, Sikh, and Muslim, but also Parsee, Anglo-Indian, and British colonial. Each story was unique, each was valuable in its own way. What emerged was a complex tapestry of the Partition 'experience'. It is not by any means definitive, perhaps it can never be complete.

When we met our interviewees so many would say 'Why would you want to hear my story? I am not Nehru, Jinnah, or Gandhi.' And yet what would come out was the most epic story, fit for a Hollywood movie. The *lived* experience. Many interviewees we spoke to were saying the words aloud for the first time. It was often raw and painful. Their children and grandchildren may have known fragments of the story, but many sat stunned in the corner, some silently shedding tears. It was such a privilege to hear these stories. There was an added poignancy. These were now elderly people talking to us, and it was the last chance for many of them to speak. They told their stories without reticence, wanting them recorded, respected, and heard. All I could think was how could these stories have been locked away for so long?

The Partition of India

What is Partition? Having spent time in the run-up to the 70th anniversary talking to programme-makers about it and seeing their blank responses, I no longer assume that even the most well-informed person knows about this part of twentieth-century history. It is a contested piece of history, though historians do agree on some facts.

In the centuries before Partition religious groups had co-existed. For example, in the regions of Bengal and Punjab – which were later to be divided – people shared a common language and culture, they celebrated each other's festivals. Ties to your land could be stronger than ties to your religion. Of course, there were differences: there was no inter-marriage and most Hindus did not eat at the homes of Muslims. There were also socioeconomic disparities, and there were cases of localised outbreaks of communal trouble prior to Partition. But, mostly, these people of different faiths had lived together, side by side, for centuries. However, religious identity in the years leading to Partition was becoming stronger. Many argue that it was stoked by the British policy of divide and rule. There were two visions of independence emerging, and the people were becoming politicised along religious lines.

The Congress Party led by Jawaharlal Nehru wanted India to remain united after the British left. The leader of the Muslim League, Mohammed Ali Jinnah, was bitterly at odds with Nehru. He felt India's 100 million Muslims – a quarter of the population – would be marginalised by the Hindu majority. He demanded safeguards – even a separate homeland for India's Muslims.

FIGURE 6.1 Direct Action Day, 16 August 1946, Calcutta, India.
Keystone/Stringer/Hulton Archive/Getty Images.

The first Partition massacre took place in Calcutta on 16 August 1946 after Jinnah called a Direct Day of Action (Figure 6.1). Four thousand people died in the ensuing violence between Hindus and Muslims. In the weeks and months that followed, communal violence spread across parts of Bengal, Bihar, the outskirts of Delhi, Rawalpindi, and then in early 1947 to the cities in the Punjab. There was a symmetry to the violence. All sides were victims, all sides aggressors.

By this time the government of Clement Atlee in Britain had already announced that British rule in India would come to an end by June 1948. Britain was bankrupt after the Second World War. It could no longer afford a presence in India. Against this backdrop of communal fighting, Lord Mountbatten, along with Jinnah, Nehru, and Baldev Singh for the Sikhs, agreed to a Partition plan on 3 June 1947. The provinces of the Punjab and Bengal were to be divided along religious lines. It was announced that it would take place in a breath-taking 10 weeks. The princely states, including Kashmir and Hyderabad, were to decide for themselves which country they would join.

Cyril Radcliffe, a British lawyer who had never been to India, arrived in Delhi in early July and stayed at the Viceroy's house surrounded by out-of-date maps and census information. In just over five weeks, Radcliffe made his decision on where the border would lie. He never once visited the areas he was partitioning. Independence and Partition – forever twinned – officially took place on 15 August 1947, though marked a day earlier in Pakistan. Astonishingly, millions had no idea which side of the border they were on, as the boundary line was not disclosed until two days later on 17 August.

Days before independence, rumours were flying around about which country cities would be going to. The prized city of Lahore went to Pakistan. Rumours swirled that Gurdaspur would be in India, but it went to Pakistan. Moga was where my dad was – his family had no idea which side of the border it would end up on and had made preparations to leave if it had gone to Pakistan – but the city went to India and they did not have to move again. The two states were divided broadly along religious lines. West Punjab – along with Sindh, Baluchistan, the North-Western Frontier Provinces, and East Bengal – formed the new Islamic state of Pakistan. East and West Pakistan were separated by nearly 1,000 miles of Indian territory (Figure 6.2). Travelling by sea between Pakistan's two major ports of Karachi and Chittagong took five days. Jinnah described his new country as a 'truncated ... moth-eaten' territory.

The flow of refugees had already begun prior to Partition. Sikhs and Hindus to India, Muslims to Pakistan. But what happened after independence, neither fledgling governments nor the colonial authorities had foreseen. Millions were on the move at the time of Partition, most for fear of being a minority in a new country. At the time of Partition, most of the refugees were from the Punjab. It is estimated between 10 and 12 million migrated in the months around Partition. The 1940s were an intense time of dislocation globally, but even so, outside war and famine this was the largest migration in human history.

By the end of 1947, the province of Punjab – which had been mixed for centuries – was virtually segregated along religious lines. Migration in Delhi and Sindh began only once Partition refugees had started arriving. By September 1947, tens of thousands of Muslims were fleeing Delhi, and at the start of 1948, Hindus from the Sindh were escaping to India.

FIGURE 6.2 Map of the British Empire in India on the eve of Partition, 1947. Universal History Archive/Contributor/Universal Images Group/Getty Images.

Refugees from both sides of the border occupied newly vacated homes, or lived in makeshift camps, and started their life over again, sometimes having to learn a new language. Long after 1947, refugees continued to move within and across new borders in response to smaller-scale communal upheavals. Partition violence continued into 1950 in Bengal, causing further movement. This finally led in 1971 to the secession of East Pakistan, forming Bangladesh and displacing almost 10 million people – some of whom are still refugees today.

The scale and intensity of Partition violence is shocking. The purpose was to eliminate the other religion: by death, by physically branding them, or by defiling their women. No one is sure how many died. Lower British estimates are of 200,000, while some South Asian scholars argue up to two million were killed. Over 75,000 women were raped or abducted, mostly by men of the 'other' religion. Some women were killed by their own families to protect them from so-called 'dishonour'. To this day, historians still debate how things could have descended to these levels of depravity. How could people turn against each other like this? This violent birth of two nations has traumatised them, so much so that it still affects the politics on the Indian subcontinent and the way people memorialise Partition. There has never been a reckoning – in India, Pakistan, or Britain – into what happened. The first museum to Partition opened only last August in Amritsar, India.

Partition Voices

The statistics of the violence are so staggering that you can lose sight of the individual story. One man I interviewed is Harchet Singh Bains, who now lives in Hitchin in Cambridgeshire. He was a Sikh living in Western Punjab in the summer of 1947, aged 11, when it became part of Pakistan. He was woken one morning and told to gather in the village square. All the villagers were there and they began to walk in a *kafila*, or caravan. On their journey, as they passed other Sikh villages, the group grew until he thinks there were thousands of them in a long line. The *kafila* was regularly attacked by Muslim mobs, and he saw many people injured, but his overriding memory is his gnawing hunger, and how he ate leaves to survive.

Harchet only knew he was safe when he was handed chickpeas in Fazilka, which he learnt was part of India. Like so many people, his family thought they were just moving temporarily, and would soon be home. Harchet's father had even left the key with a Muslim neighbour, giving him instructions to look after the animals until their return. But Harchet never did go back to the place he grew up.

Harchet is one of many South Asians who came to Britain in the early 1950s and has lived here ever since. South Asians have a long history in Britain pre-dating Partition. There has been a flow of princes, students, lawyers, servants, and seamen stretching back to the seventeenth century and earlier. But post-war migration assumed a volume and scale not seen before. And there is a link between Partition and migration to Britain. The areas of Indian and Pakistani Punjab and Pakistan-administered Kashmir which were greatly disturbed by Partition were major contributors to the emigrant flow to Britain. The region of Sylhet (then East Pakistan, now Bangladesh) also had a tradition of emigration over the northern border into the Indian state of Assam, a flow that was reversed after Partition. 'As the British tide retreated from India, the Indian, Pakistani and then Bangladeshi waters disturbed by Partition lapped into Britain,' notes the Oxford geographer Ceri Peach.[1] Crucially these were also groups of people who had networks and connections prior to Partition to particular parts of Britain.

South Asian migration started seriously in the 1950s. The British Nationality Act of 1948 conferred British citizenship on all who lived in the British Empire and Commonwealth. A citizen of India and Pakistan was a citizen of the United Kingdom. Limitations began to be placed on this principle only in 1962. Post-war migration was essentially related to British labour shortages in the mills, factories, foundries, and public services. South Asians, and others from across the former British Empire, came in their thousands in search of a better life. The end of empire therefore caused huge migration flows – not only on the Indian subcontinent, but also between South Asia and Britain itself. Our histories – those of the Indian subcontinent and Britain – are inter-connected. We need to understand this past.

The 70th anniversary of independence and Partition in August 2017 was a watershed moment. Something happened. There was a huge volume of coverage: *Newsnight* devoted an entire programme to it; BBC One commissioned a documentary following four families returning to the subcontinent to explore their Partition past; Salman Rushdie's *Midnight's Children* was serialised by BBC Radio 4; the BBC *News at Ten* co-presented its programmes from Lahore and Amritsar; the *Guardian* had an excellent series of reports; and the *Telegraph* published an article entitled 'Everything you need to know about Partition'. I had researched

---

[1] Peach, C. (2006). 'South Asian migration and settlement in Great Britain, 1951–2001'. *Contemporary South Asia*, 15(2), 133–46, p. 134.

the coverage for the previous anniversaries, and this was by far the most extensive, sensitive, and thoughtful and was placed prominently in the schedules and newspapers.

Most British people who consumed this media coverage were shocked at what happened. They were shocked that they had been ignorant of the scale and fervour of the violence, and also at the shambolic nature of the British departure. Some South Asians responded by saying they had no idea of this history either, and it prompted many questions and conversations amongst family members. One woman got in touch on Twitter to say her 89-year-old father shared what she called his 'awful story' for the first time upon hearing the *Partition Voices* radio programmes. The media coverage had been a catalyst for him to talk about his own painful experiences. ('How he kept quiet for 70 years I do not know,' the daughter added.) The coverage also did something else – it showed all sides in the history of Partition: Hindu, Muslim, Sikh, and British. One person who wrote to the BBC said that they had had 'no idea the other side suffered too'.

Remembering Partition

But why this coverage now? Why are people finally opening up in a way they had not 70 years before? I think there are a number of reasons. This is probably the last big anniversary where the people who lived through it will still be alive to talk about it. I think the South Asian diaspora are curious about their history and how it informs their identity. As Brexit looms, and the country is going through huge changes, and also questioning its international standing, I do wonder whether there is some curiosity about why and how Britain's last big retreat from Empire happened. But I also think that finally, and perhaps most importantly, we are asking and listening in a way we had not before. And it is about time. A public space was finally created where people could discuss their stories. Oral history projects are now taking place across Britain in local areas, and there is a hunger to record these stories. It is a race against time.

There has been silence not only in Britain, but also in India and Pakistan. But there are different reasons for the silence here. Many who came to Britain in the years after Partition were just getting on with their lives, fighting different battles to be accepted in this country. There was no time to dwell on the past. And, unlike back home, they were fighting together – Indians and Pakistanis – whether it was within the Indian Workers Association in Britain, or in the fight against racism. In Britain, they were all lumped together as 'Asians'. No one saw the differences so exposed on the Indian subcontinent. Their children and grandchildren knew little of life back home, so why talk of it? There was an institutional silence too – no one talks of Partition and the ensuing migration and violence, end of empire, or even empire in British schools. So there was no public space to discuss it. And, for my father, it was simple: why would you want to burden your children with this history? But there is another reason too. In Partition there were perpetrators on all sides, and obviously within families. How do you begin to talk about that? So for all these many reasons there was silence.

Yet even silence can be noisy. And there were signs. The unity between different South Asian groups was fragile and had dissipated by the early 1980s. Prejudices continued – 'Don't marry a Muslim or Hindu.' But there were other clues too – when old friends got together they would discuss and reminisce about 'home'. But 'home' may not be the country that the extended family now lived in on the Indian subcontinent. 'Home' could be the so-called 'enemy' state. For the subsequent generations in Britain, where identity is complex, conversations were beginning. Silence was breaking.

I want to introduce you to Iftkhar Ahmed, known as John (Figure 6.3). He has lived in Brighton since 1951. Inside his very ordinary Victorian home he told me an extraordinary story of his escape from Delhi to Lahore, one that his son Nick, who is now in his fifties, had only recently heard of.

Iftkhar's hair-raising journey culminates in him arriving, alone, aged 15, at Lahore train station in September 1947. He falls asleep on the platform, and when he wakes he feels wet. Around him are people. But they are all dead. He learns that while he was asleep a train arrived from India with everyone on board (except the driver) murdered. The bodies were then loaded onto the platform. The wet he had felt that morning was blood.

FIGURE 6.3 Photograph of Iftkhar Ahmed. Courtesy of the BBC.

Iftkhar is, with the help of his son, now writing the story of his life: his birth in British India, his flight to the new country of Pakistan, and then his migration to Britain – the home of his former colonial ruler, a place where he has spent most of his life, where his children and grandchildren were born.

Why does it matter that we collect these stories of double migration in Britain? Some will question the trustworthiness of accounts given 70 years after the fact. Urvashi Butalia, one of the foremost oral history chroniclers of Partition in India, raised this question at the beginning of her seminal book *The Other Side of Silence*:

> I have come to believe that there is no way we can begin to understand what Partition was about unless we look at how people remember it. I do not wish here to carry out a literal exercise of Partition and then attempt to penetrate their narrative for its underlying facts to arrive at an approximation of some kind of 'truth'. Instead I wish to look at the memories themselves – even if they are shifting, changing, unreliable.

She quotes the Holocaust historian James Young, who writes 'Whatever "fictions" emerge in the survivors' account are not deviations from the "truth" but are part of truth in that particular version.'[2]

We would never attempt to understand the Holocaust, for example, without hearing the testimony of survivors. We have scant accounts of the experiences of First World War soldiers, because of their trauma, which is a gap not only for families but also in the historical archive. These first-hand accounts are just one important facet of the Partition story; 70 years on, memory may have changed, but how it is remembered today is relevant. These stories have a particular resonance in this country in helping form the identity of second- and third-generation South Asians. These stories can show the connections rather than the differences, which can be missing from other more formal documentary evidence on the subject, or when Partition is viewed through the prism of 'high politics'. The lived stories tell you so much.

Raj Daswani's story is one of love as well as loss. The 15-year-old Raj wanted to marry his neighbour, Yasmin. He was a Hindu, she a Muslim. At that time, inter-marriage was unheard of. It is a real Romeo and Juliet story. The two would sit on the balcony together and imagine spending their lives together. They would look up at the moon, and Raj would tell her 'One day I shall give you this moon.' Raj was from Karachi, in Sindh Province, which after Partition would become part of Pakistan. Here, his Hindu family lived very closely with their Muslim neighbours. When Raj and his family left, all their Muslim neighbours came out crying, begging them to stay, saying they would protect them. Leaving Yasmin, seeing her for the very last time, has stayed with him for 70 years. They never saw each other again.

And for all these years since, not only had he thought about what became of Yasmin, but he also had a profound yearning for Sindh. He still speaks the language at home. He identifies himself first and foremost as being from there. His children and grandchildren were born in Britain, but he and his wife still teach them Sindhi. It is even more important to do, he says, as Hindus no longer have a presence there.

---

[2] Butalia, U. (2000). *The Other Side of Silence: Voices from the Partition of India.* Durham: Duke University Press, pp. 13–14.

Legacies of Partition

In Britain these stories matter. They matter for British South Asians in explaining who they are. These stories are part of their identity. These accounts also show that all sides suffered. It is important that not just the stories of violence by one side are heard, which could foster further division within the South Asian community today. Rather, we need to recover the stories of co-existence and shared humanity; the affinity to a land lost, but also the acts of kindness that were shown during this terrible time. Finally, it is important too that the wider population understand why South Asians are in Britain, and the connection to Empire. This is a British history, not just a niche South Asian history. It is just as important as learning about the Tudors and the Norman Conquest. It explains why Britain looks the way it does today.

It is hard to sum up the interviews that I conducted – each experience was unique. What they did share was a profound sense of loss. Loss of a happier past, loss of home, loss of homeland, loss of friendship, loss of security, and grief. Many said they wanted to return to the place of their birth one more time.

Notions of home are not so simple. My father is a good example. Ask him what home is and he will say Britain as it is where his children and grandchildren are, India as his parents lived there, and Lahore as it was his birthplace. I watched the BBC Two programme *Dangerous Borders* during the Partition season with him. A young second-generation British Pakistani was in Lahore exploring his roots. He was at a bazaar, and I thought that this must be so familiar to my father, yet Lahore is frozen in time for him.

My dad wanted to go back, but never did. But others did go back and visit. Their accounts are incredibly moving. Mohindra Dhall, who lives in Edinburgh, returned from Pakistan with a rose-coloured brick. The brick was from the home he was bought up in near Lyallpur, now Faisalabad, in Pakistan. Why? Because he wanted to remember the place he was forced to leave, on his sixth birthday, on 3 September 1947. He took it all the way back to Edinburgh, where it sits in a glass cabinet in the centre of his living room. There's no missing it. It starts many conversations. He is the only one from his Hindu family who has ever returned to the subcontinent.

FIGURE 6.4 Photograph of Raj Daswani. Courtesy of the BBC.

Raj – who left his first love Yasmin – feels an equally strong connection to Sindh province. On arriving at Karachi airport, the first time he returned to Pakistan many decades after he had fled, Raj kissed the dust from the ground and put it on his forehead. His words on arriving were 'Mother, I have returned home.' He took stones from the earth back to his home in North London. He says it is so he can be connected to his soil (Figure 6.4).

Tara, a young British Asian from Birmingham, discovered a photograph of her grandfather which led to her and her family unearthing his past. He was pictured at the Viceroy's House, where he worked closely with Lord Mountbatten. They found out he was born in Pakistan. The only thing he kept from that time was a tile from their Pakistani home that is now part of his house in Delhi. Her grandfather passed away, and the family, now spread out across the world, want to keep that tile – as it is all they have from that time.

Soil, bricks, stones, a tile. These people were lucky to have a tangible reminder of what was left behind. It is touching, and there is something very poignant in elderly people, or their children and grandchildren, holding onto whatever scraps they can of a former life. In many cases – even though you may have migrated once back in 1947, and then migrated to Britain – that belonging, that connection to the place of your birth, endures. It goes back to a time before division, borders, and partition. A time of shared humanity – and what lasts and still lasts is language and culture: be it food, music, or traditions.

When Poonam Joshi's mother died, she wanted an Urdu song from her youth played at her Hindu funeral. Poonam wanted her mother's ashes scattered in Lahore as it was the place she was the happiest – yet it was a place her mother had left in 1947 and not returned to since. Yet the pull of Lahore shaped her life.

This need for connection comes from something very deep. It is a statement that 'I lived there once too. It is also my land, even if I have migrated once and then once more.' Punjab now is virtually segregated. Bangladesh has a Hindu population of around 8%. India has between 170 and 180 million Muslims, but Pakistan, after its separation from Bangladesh, has only around 2% Hindus and Sikhs. These shifts of people – whether to India, Pakistan, or Britain – carry stories, memories of a time when people managed to live together. These were the stories people wanted to remember and share too. If there is a purpose to oral history, it is that these stories – not just the ones of horror and violence – survive down the generations. These stories are testimony that there were once Hindus and Sikhs in Lahore, Muslims in Amritsar. And 70 years on, these people still feel they belong to the place they had no choice but to leave – even if it is 'enemy' land. Borders were put up, migration happened, but the idea of home can still remain in the places left behind. Will that carry on with the next generation? Some in the second and third generation in Britain are trying to piece the stories of their parents or grandparents together.

My father did end up telling me his stories, what he had seen. It was pretty awful and I understood why he would have preferred silence. But something else also occurred to me. I had always wondered how he could have left India to live in England in the 1950s on his own. But then I realised, the worst had already happened to him. The original migration following Partition was so difficult, and wrenching, that the second migration to Britain could never have been as dislocating as the first one.

These are complex stories of loss, displacement, belonging, and – for the second and third generation making their way in Britain – it is part of who they are. But for British people it is important that they understand their South Asian compatriots: why they are here, how their presence in significant numbers came to be in the early 1950s. This is essentially a story of Empire, and how it ended. Yet this story is not taught in schools. It is not taught in the way that Black History or the Holocaust rightly are. Yet it is crucial to explain the South Asian presence, and important for social cohesion. It also matters for South Asian groups – to understand each other.

The migration that followed Partition still has ramifications on the Indian subcontinent today. It still tangibly affects lives amongst the diaspora in Britain. And yet we are only just learning about it after all this time. The threat of Partition, and then Partition itself, sparked the epic tide of humanity who were on the move in August 1947. Partition also caused many who had suffered to migrate to Britain in the years that followed. Partition was devastating for so many, many families. Even 70 years on, people across Britain are still only beginning to come to terms with what it means to lose your home all those years ago. Next time you see an elderly Indian, Pakistani, or Bangladeshi, ask them about 70 years ago, and you may hear an extraordinary story.

Further Reading

Ansari, S. (2005). *Life after Partition: Migration, Community and Strife in Sindh: 1947–1962.* Oxford: Oxford University Press.

Ballard, R. (ed.) (1994). *Desh Pardesh: The South Asian Presence in Britain.* London: Hurst and Co.

Butalia, U. (2000). *The Other Side of Silence: Voices from the Partition of India.* Durham: Duke University Press.

Chatterji, J. (2007). *The Spoils of Partition: Bengal and India, 1947–1967.* Cambridge: Cambridge University Press.

Chatterji, J., and Washbrook, D. (eds.) (2013). *Routledge Handbook of the South Asian Diaspora.* London: Routledge.

Hajara, N. (2015). *Midnight's Furies: The Deadly Legacy of India's Partition.* Stroud: Amberley Books.

Khan, Y. (2007). *The Great Partition: The Making of India and Pakistan.* New Haven: Yale University Press.

Puri, K. (2019). *Partition Voices: Untold British Stories.* London: Bloomsbury.

Talbot, I., and Singh G. (2014). *The Partition of India.* Cambridge: Cambridge University Press.

Vizram, R. (2002). *Asians in Britain: 400 Years of History.* London: Pluto Press.

Von Tunzelmann, A. (2007) *Indian Summer: The Secret History of the End of an Empire.* London: Simon & Schuster.

7 Migration in Science

VENKI RAMAKRISHNAN

Science is an international endeavour. Its development depends on exchanges of ideas and expertise, which are made possible by people moving from one part of the world to another. I myself am an immigrant at least twice over. I grew up in India and studied in the United States, spending almost three decades there before moving again to England almost 19 years ago to work at the Medical Research Council (MRC) Laboratory of Molecular Biology (LMB) in Cambridge. I was not the first immigrant to arrive there. In fact, the first three directors of the LMB were all immigrants. The first director, Max Perutz (1914–2002), was Austrian. The second, Sydney Brenner (1927–2019), and the third, Aaron Klug (1926–2018), were both South Africans. The current director, Jan Löwe, is also an immigrant, from Germany.

As I walk past the large foyer in the new LMB building to go up the stairs to my lab, I pass a poster that lists all the Nobel laureates of the LMB (Figure 7.1). You can see the lab has been incredibly successful; there have been 17 Nobel laureates to date. Well, there are actually only 16 individuals because Fred Sanger appears twice – he is one of the few people to have won two Nobel Prizes. When you look at these people, it is striking that well over half of them were immigrants to England. It just shows you the world today, where scientists come from many different countries to work in the big centres like the LMB to make their contributions. Moving to a more national level, there is a similar story at the Royal Society. Three of the last five Presidents of the Royal Society were also immigrants to Britain, and a sixth, Michael Atiyah, was the son of a Lebanese immigrant.

Clearly, migration is important for science today, particularly in Britain. But has migration mattered for science in the past? In this chapter,

FIGURE 7.1 Poster of Nobel laureates associated with the MRC Laboratory of Molecular Biology. Photograph courtesy of Venki Ramakrishnan.

I will give you some examples of how ideas travelled and developed as a result of migration. I will conclude with a personal reflection on my own journey as a migrating scientist.

Arabic Science

Try and multiply these two numbers: CIX and XCVII. What you'll notice is that it is very difficult because they are expressed as Roman numerals. These are the numbers that existed in Europe for almost a millennium, from at least 500 BC to 1400 AD. However, if you convert these numbers to what we call Arabic numerals, then suddenly the multiplication becomes much easier: $109 \times 97 = 10,573$. The beauty of these Arabic numerals is that, unlike the Roman notation, each number has a quantity but it also has a place value. You have units, tens, and hundreds, and so on. The other interesting thing is that Arabic numerals

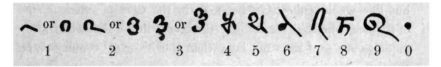

FIGURE 7.2 Numbers from the Bakhshālī manuscript, as listed in Hoernle, R. (1887). *On the Bakhshālī Manuscript.* Vienna: Alfred Hölder.

have a special number zero which signifies nothing but also acts as a placeholder. Without zero, and a regular assignment of digits, it was almost impossible to do long multiplication in Roman numerals.

Our modern positional notation for denoting numbers has its origins in India a couple of thousand years ago and such numbers are thus more properly referred to as Hindu–Arabic numerals. One of the oldest examples we have of Arabic numerals comes from the Bakhshālī manuscript held in the Bodleian library, which is written in Sanskrit. Parts of the Bakhshālī manuscript have recently been carbon-dated to between 220 and 380 AD. The manuscript itself is a compilation of mathematical rules in verse with prose commentary. They describe algorithms for solving a variety of problems, including linear equations, quadratic equations, arithmetic series, and so on. If you look closely at the manuscript, you can see the occasional dot (Figure 7.2). These dots are the oldest surviving example of the symbol for zero. If you look further in the manuscript, you can see the other numbers represented in the Sanskrit script, with different symbols for each of the 10 digits. Many of these, you'll notice, are not so dissimilar from the modern numbers we use today.

These numerals, and the accompanying positional notation, then spread far beyond India, largely through the influence of Islam. The different waves of Islamic conquest began in Saudi Arabia in the seventh century. Islam then spread throughout India and North Africa, and into Europe, reaching as far as Spain and southern France by the eighth century. During this period of conquest, scientific ideas – including positional notation and the symbol for zero – migrated from India, through the Islamic world, and into Europe.

Arabic numerals, as we've seen, are much easier to manipulate than Roman numerals. They allowed mathematicians in the Islamic world to develop a number of new and important techniques. Most significantly, the concepts of algebra and an algorithm were both invented in the mediaeval Islamic world. Indeed, you can identify the Arabic origins of these words from the prefix *al-* (simply meaning 'the' in Arabic). The word algorithm is named after an eighth-century Persian mathematician named Muhammad ibn Musa al-Khwarizmi (*c.* 780–*c.* 850). (Mediaeval European scholars Latinised al-Khwarizmi's name to 'Algorithmi'.) He also invented the concept of algebra, which comes from the Arabic *al-jabr*, meaning 'the reunion of broken parts'.

Along with mathematics, the Arabs also advanced the science of cryptography. Prior to the ninth century, almost everyone believed it was impossible to break what is known as a substitution cipher. This is where you simply substitute one letter of the alphabet for some other unrelated letter. However, the substitution wasn't completely random. Substitution ciphers were in fact based on what is known as a key phrase. For example, if your key phrase is 'Darwin Lecture', what you would do is remove the space and any repeated letters. Here, R and E are repeated, so you would only use their first occurrence, resulting in the key phrase 'DarwinLectu'. Next, you would put this key phrase underneath the alphabet, running from A to Z. Then at the end of the key phrase you would start off with the next letter in the alphabet (in this case V) and keep going with all the letters that hadn't been used in the key phrase. When you get to Z you'd go back to A. (In this case, A is part of the key phrase, so you would actually use B, and so on.) You then substitute the letters in the message you want to encode with the letters from the cipher (Figure 7.3). The point of the key phrase was to allow people to easily remember the substitution cipher without writing it down. And because of this, the substitution cipher was considered uncrackable.

Of course, letters are never completely random. My wife and I like to play Scrabble, which has a particular distribution of letters. There are far

abcdefghijklmnopqrstuvwxyz
darwinlectuvxyzbfghjkmopqs

FIGURE 7.3 Substitution cipher for key phrase 'Darwin Lecture'.

more tiles of the letter E than of B or C. And there is only one each of Q and Z. (Foreign-language editions of Scrabble have different distributions of letters.) How did this distribution come about? Alfred Mosher Butts, the person who invented Scrabble, just took a newspaper article and counted the number of times a particular letter occurred. He then assigned that fraction, rounded up to the nearest percentage, to the 100 tiles, ensuring that there was at least one of each letter. So he assigned 12% of letters to E, 2% to B, 1% to Q, and so on.

What the Arabs realised is that you could take a completely unknown encoded text and simply count how often each character appeared. This solution was first proposed by a mathematician born in Baghdad in the ninth century named al-Kindi (*c.* 801–*c.* 873). Depending on the language, different letters will have higher frequencies. If it is an English text, for example, you know the most frequent letter is likely to be an E. If not that, then it is most likely to be an A or an I. Checking for the most frequent letters, al-Kindi was then able to guess common words to obtain more of the substitution code and gradually decode an entire message. This was the first time the substitution cipher had been cracked.

At first, this algorithm of using frequency analysis for solving the substitution cipher wasn't known to everybody. But, by the sixteenth century, knowledge of it had reached a few people in England. As described by Simon Singh in the dramatic opening to his *The Code Book*, Mary Queen of Scots was leading the Catholic rebellion against Queen Elizabeth. She had been imprisoned, and there was strong pressure on Elizabeth to execute her. However, Elizabeth, because Mary was her cousin, did not want to do so unless there was actual proof of treason. Elizabeth's counsellor, Francis Walsingham, was a well-informed spymaster who knew about Arabic cryptography through his contacts in Spain. So he knew that it was possible to crack a substitution cipher. As a result, he was able to crack Mary's communications with her supporters, which were written in code. It soon became very clear that she was giving direct approval to her supporters' resistance against Elizabeth. This effectively sealed her fate and she was executed on 8 February 1587. Ultimately, this is a case where the migration of science, from ancient India, through medieval Islam, and on to early modern Europe, had a major historical consequence.

Indian Science and the British Empire

So far I've explored the ancient and mediaeval world of the Islamic conquest and the migration of knowledge and ideas from East to West. Next, I'd like to turn to the period between around 1750 and 1950, with the rise and fall of the British Empire in India. In 1841, the Parsi engineer and shipbuilder Ardaseer Cursetjee Wadia (1808–1877) became the first Indian to be elected as a Fellow of the Royal Society. However, for the next 77 years, no further Indians were made Fellows of the Royal Society. This was partly because there were no institutions of higher learning in India and very few Indians could afford to go to Britain to study. However, in 1882 four universities and 67 associated colleges were established in India. This marked the beginning of a new wave of Indian science, that was initially catalysed by a flow of knowledge from Britain to India, but that in turn influenced European mathematics and physics.

In my office at the Royal Society I have a bust of the famous Indian mathematician Srinivasa Ramanujan (Figure 7.4), who was born in 1887. Ramanujan grew up in South India, spending most of his adult life working in Madras. He was a self-taught mathematical genius. In 1913, whilst working as a clerk at the Port of Madras, Ramanujan wrote to G. H. Hardy at Trinity College, Cambridge, setting out some of the mathematical theorems he had discovered. These included work on prime numbers, theta functions, continued fractions, and partition formulae. Hardy was impressed, and, after making enquiries, arranged for Ramanujan to come to Cambridge. He was elected a Fellow of the Royal Society in 1918 and then a Fellow of Trinity College. Ramanujan's recognition in Britain gave Indians enormous self-confidence. Two centuries of colonisation had inculcated in Indians a feeling that they were somehow inferior to Europeans. And, unlike Wadia, Ramanujan was neither Anglicised nor wealthy. He had a strong Indian accent and came from a fairly poor family – and yet got the highest recognition. Even today, his notebooks continue to be studied by mathematicians.

This really spurred on Indians to become very ambitious in doing science. Chandrashekhara Venkata Raman (1888–1970) won the Nobel Prize in Physics in 1930 following his discovery of Raman scattering. Jagadish Chandra Bose (1858–1937) is credited, among many other

FIGURE 7.4 Bronze bust of Srinivasa Ramanujan. Reproduced with permission of the Royal Society and the Estate of Paul Granlund.

things, with co-inventing radio transmission. Not long afterwards, Satyendra Nath Bose (1894–1974), helped to lay the foundation for the behaviour of an entire class of subatomic particles, which are collectively known known as 'bosons', of which the Higgs boson is a famous recent example. And Homi Jehangir Bhabha (1909–66) made major contributions to the field of nuclear physics in the 1940s and 1950s. Unlike earlier generations, all of these individuals were elected Fellows of the Royal Society.

Now I want to talk a bit more about a second Indian scientist, named Subrahmanyan Chandrasekhar (1910–95) (Figure 7.5). In fact, it was Chandrasekhar who donated the bust of Ramanujan to the Royal Society. Chandrasekhar was educated at Presidency College in Madras, one of the

FIGURE 7.5 Subrahmanyan Chandrasekhar, of the Yerkes Observatory at the University of Chicago. Bettmann/Contributor/Getty Images.

four universities that the British founded in the late 1800s, and, at the age of 19, he wrote his first paper, entitled 'The Compton scattering and the new statistics'. Following this, Chandrasekhar was accepted as a research student at Trinity College, Cambridge.

In his first migration, Chandrasekhar left India for Cambridge in 1930. In those days, it took a few weeks to arrive in England by steamship. During that long journey, Chandrasekhar continued to do physics. He realised that, if stars had a mass larger than a certain limit, the force of gravity would be so strong that it would overcome any hydrostatic pressure, and they would collapse into themselves. That mass limit is now known as the Chandrasekhar limit, and this idea of collapse is the basis of black holes, which were discovered about 40 years later.

When Chandrasekhar arrived in Cambridge, he touted his theory to his colleagues, and many of them thought it was a brilliant idea. But one of them was almost vitriolically opposed to it. Unfortunately for Chandrasekhar, that man, Arthur Eddington (1882–1944), was the most powerful and famous astrophysicist of his time. Eddington then did something that was rather underhand. He invited Chandrasekhar to present his findings at a meeting of the Astronomical Society in London on 11 January 1935. During the weeks that led up to this lecture Eddington would constantly ask Chandrasekhar about his ideas without telling him that he was going to give his own lecture immediately afterwards. And so, after Chandrasekhar had spoken, Eddington essentially ambushed him in his own lecture, in which he completely dismissed the young man's ideas as a mathematical anomaly having no relationship whatsoever to physical reality.

Despite this severe psychological setback, Chandrasekhar went on to do good work, and, to Trinity College's credit, they elected him to a Prize Fellowship, despite Eddington's scathing assessment. But Chandrasekhar felt that if he stayed on in England he would always be working under Eddington's shadow, so in 1936 he left to work at the University of Chicago. However, he didn't actually work at the main campus of the University of Chicago for a very long time. His first years were spent at the Yerkes Observatory in Wisconsin, about a two-hour drive from the university. During this time, Chandrasekhar would drive once a week to teach a course in advanced physics, that on one occasion had just two

Chinese graduate students. He was obviously an excellent teacher, because his entire class, consisting of Tsung-Dao Lee and Chen-Ning Yang, was jointly awarded the Nobel Prize in Physics in 1957 for their work on the non-conservation of parity in nature.

Chandrasekhar went on to achieve considerable recognition as a world-famous astrophysicist. He was elected a Fellow of the Royal Society in 1944, only seven years after his arrival in Chicago. However, he had a rather long wait for the Nobel Prize because it is not awarded for a theory unless there is experimental proof of it. It took about 40 years to verify that black holes were real before Chandrasekhar was finally awarded the Nobel Prize in Physics in 1983. This was effectively for work he had done at the age of 20, beginning in India. But, to get to that point, Chandrasekhar had been on a journey from India, through Britain, and – after experiencing the hostility of Eddington – on to the United States.

Jewish Science in the Twentieth Century

The next history of migration I'd like to discuss is that of Jewish scientists in the twentieth century. To understand this history, we need to consider the context of the 1930s and the growth of fascism and antisemitism. This too led to another kind of migration, much of it forced.

Lise Meitner (1878–1968) was born in Vienna to Jewish parents (Figure 7.6). She is possibly the most important example of a woman who deserved the Nobel Prize but was not awarded it. After studying physics at the University of Vienna, Meitner moved to Germany to do research under Max Planck (1858–1947). She later became a Professor of Physics at the University of Berlin. Meitner's Austrian citizenship initially protected her against the widespread removal of Jews from academic positions after the Nazis' coming to power in 1933. However, in July 1938, shortly after the incorporation of Austria into Germany, Meitner was forced to give up her post and flee. She escaped to the Netherlands and thence to Sweden, where she did most of her work on nuclear fission.

As all this was happening, it became increasingly clear in Britain that Jewish scientists in continental Europe were in real jeopardy with the

FIGURE 7.6 Lise Meitner lecturing in Sweden in 1949. Bettmann/
Contributor/Getty Images.

spread of fascism. In response, a group of British scientists, led by the politician William Beveridge, helped set up the Academic Assistance Council in 1933. It was a charity, not run by British Jews for German Jews, but rather by British scholars committed to the value of scientific endeavour regardless of nationality, race, or creed. It was a remarkable decision by the British academic community to mount this totally selfless rescue operation.

As a result, many people, including Hans Krebs (1900–81) and Ernst Chain (1906–79), were able to come to Britain and escape the Nazis. Seventy-four of these displaced scholars became Fellows of the Royal Society, of whom six became foreign members, 16 were Nobel laureates, and 18 were knighted. As is often the case, they were hugely successful and contributed hugely to the scientific life of Britain.

The same is true in the United States. Following the rise of Nazi Germany, many Jewish scientists, including Albert Einstein (1879–1955), Edward Teller (1908–2003), and Johann von Neumann (1903–57), left Europe for the United States. Indeed, many prominent scientists in the United States today are Jewish and over 20% of all Nobel Prize winners are Jewish. Most of these are immigrants to the United States, or children of immigrants.[1]

A Personal Migration History

In the final part of this lecture, I will briefly talk about my own migration. I was born in Chidambaram, South India in 1952. When I was three years old, my parents moved to Baroda in the state of Gujarat in western India. It was almost like going to a foreign country. Everyone spoke a different language, and my earliest memories are of standing in a playground and not understanding what any of the other children were saying.

I obtained a degree in physics from the Maharaja Sayajirao University in my home town of Baroda when I was 19, and decided to go to graduate school in the United States. By my generation, the United States had

[1] 'A remarkable week for Jewish Nobel Prize winners', *Jewish Chronicle*, 10 October 2013, www.thejc.com/news/world/a-remarkable-week-for-jewish-nobel-prize-winners-1.49544 (accessed 14 October 2019).

become the default destination for postgraduate studies, and I began my PhD work in physics at Ohio University. However, as I was getting my PhD in physics I realised that physics was actually very hard. I also realised that there had been very few breakthroughs in the previous several decades that I thought of as really fundamental. I felt that if I continued in physics I would spend the rest of my life doing boring calculations that would result in publications but that wouldn't actually improve our understanding of the world.

On the other hand, there was biology. I used to and still do subscribe to *Scientific American.* In the 1970s, almost every issue of *Scientific American* reported major breakthroughs in biology. I knew that many physicists had made that transition, for example Max Delbrück (1906–81), Francis Crick (1916–2004), and Max Perutz. I thought of doing a postdoc in biology, but felt I didn't know enough biology and needed to acquire that background first. So although I already had a PhD in physics, I ended up going backwards in order to change fields and began graduate school all over again – this time in biology – at the University of California in San Diego.

After two years, I felt I had acquired enough background in biology and felt ready to do a postdoc. I came across an article on studying something called the ribosome by using a physical technique called neutron scattering. I wrote to the authors of the article, and one of them, Peter Moore, offered me a postdoctoral fellowship. Moore worked at Yale, so we ended up crossing the country yet again. This was with two young children in tow, so I have to acknowledge the fact that I had a very supportive wife who was rather tolerant of these frequent moves.

The ribosome is related to the problem of proteins and how proteins are made. Proteins are long folded chains of amino acids. The order of the amino acids in a particular protein chain determines how each protein folds up into its unique three-dimensional structure. There are thousands of proteins and they carry out all of the thousands of functions in our cells. For example, collagen, a long, three-stranded filament, makes up our skin and connective tissues; haemoglobin carries oxygen in our blood from our lungs to our tissue; and rhodopsin sits in the membranes of cells in our retinas and detects light, enabling us to see. Thus, proteins really are what makes life happen.

Proteins are made using information in our genes, which are simply stretches of DNA. Those stretches are copied into a molecule called RNA, and the RNA is then read by adaptor molecules called tRNAs (which are another type of RNA), which recognise groups of three bases in the RNA and then bring along an amino acid. The amino acids brought by the tRNAs are then connected up to make a protein, but this complicated process doesn't happen by itself. Cell biologists realised that all of these proteins are made in particles in the cell called ribosomes. The ribosome has slots for the binding of the tRNAs which bring the amino acids. When the first one has been bound in the central slot, a second tRNA can come in with its amino acid into the adjacent slot and the ribosome links up the first and second amino acids. As the ribosome moves along the genetic message on the mRNA, the tRNAs move from one slot to the next, so that new tRNAs enter the vacant slot in the ribosome and the ribosome adds its amino acid to the growing protein chain. In this way, the ribosome makes a protein as specified by the code in the gene.

The ribosome is enormous by molecular standards. Even the simplest form of it, in bacteria, consists of about 50 proteins and large pieces of RNA, which contain almost half a million atoms in total. Because it uses energy, moves and makes a product, it is often referred to as a molecular machine. Because of its complexity, understanding the ribosome has taken decades and work is still continuing. I began my postdoctoral work by using neutron scattering to tackle the ribosome, but the technique turns out not to be that useful in biology. After several years of this sort of work at Brookhaven National Laboratory, I was asked by my department what I would do if they gave me tenure. I replied, 'Well, the first thing I'll do is stop whatever I'm doing and go away on sabbatical to learn crystallography.' Luckily for me, they bought it and let me go on sabbatical.

And so I spent a sabbatical year in 1991–2 in Cambridge at the MRC LMB. I'd written to Aaron Klug, who was a Director of the LMB at the time. I went there to learn crystallography, which could be used to identify the structure of macromolecules in atomic detail. I then went back to Brookhaven. However, the lab had changed by then. It was run by the Department of Energy, which preferred to support big projects like accelerators and reactors rather than individual investigators. In 1995, I therefore decided to move across the country yet again, this time to the University of Utah, which, apart from its spectacular geography, had a fine reputation in genetics and biology in general.

When I arrived in Utah, I decided I wanted to work on the whole ribosome. However, I didn't have any idea how long it would take. A group in Germany, led by Ada Yonath, had been working on it for almost 20 years by that time and hadn't really been able to make a breakthrough. I soon realised that this was the sort of research problem where I would need stable funding. If you're on grants, and at the end of three or four years you haven't cracked the problem, they're not going to renew your grant. I also thought I would benefit from expert crystallographic colleagues. After all, I had only just learned crystallography and this was one of the toughest problems in the field.

As a result, in 1999, I left Utah and moved to the LMB, where I had spent a sabbatical year only seven years earlier. The LMB is an institution with a long tradition of tackling hard problems in molecular biology, with an emphasis on the importance of problems and not on how many papers you churn out. That's part of the reason for its success. After all, Fred Sanger wrote only about 40 papers, and he has an h-index of 18. That wouldn't get him tenure in most top universities today, but it was enough to get him two Nobel Prizes.

Anyway, the gamble paid off. In 1999, we solved first the structure of the small subunit and then that of its complex with many antibiotics. Then, a few years later, in 2007, we solved the structure of the entire ribosome – about half a million atoms – with the transfer RNAs and a piece of the genetic message in place (Figure 7.7) As a result of the structural work and lots of biochemical work by people over many decades, we now understand roughly how the ribosome works.

Broadly, the two subunits work together, reading the genetic code and joining up the amino acids into a chain. This chain then comes out through a tunnel on the other side of the ribosome. Eventually, the ribosome reaches the end of the gene. When that happens, a special protein called a termination factor binds to the ribosome and then cleaves off that newly made protein so that it can go off and do its thing. And then two other proteins bind the ribosome and split it apart so that the whole process can start all over again. This all happens incredibly quickly. In fact, while you've been reading this, tens of thousands of ribosomes in all your cells have been churning out thousands and thousands of copies of all sorts of proteins.

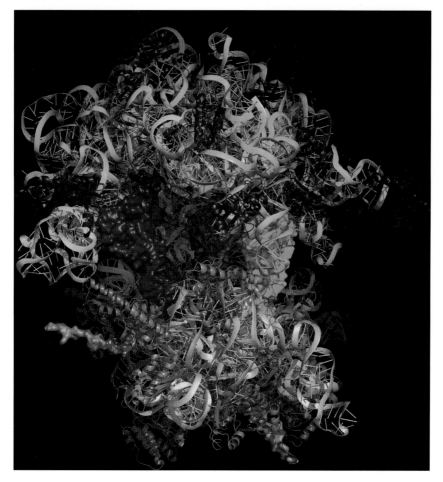

FIGURE 7.7 Three-dimensional computer model of the structure of the ribosome.

Science and Migration Today

Today in Britain, 30% of researchers are immigrants, and about half of those are from the European Union.[2] Immigrants are often vastly over-represented in the highest echelons of science. In the United States, about

[2] Frenk, C., Hunt, T., Partridge, L., Thornton, J., and Wyatt, T. (2016). *UK Research and the European Union: The Role of the EU in International Research Collaboration and Research Mobility.* London: The Royal Society.

25% of National Academy members are immigrants. A similar fraction of Nobel Prize winners are immigrants.[3]

The question then is why is migration important in science? I think there are two answers to this. First, migration allows the exchange of new ideas and expertise and new ways of doing things. People who move from elsewhere bring new ideas, new techniques, and different ways of thinking about problems. Second, immigrants often tend to overachieve. In order to be selected to come to a place like Cambridge or Harvard, they have to be better than average to begin with. But when they arrive at a new and unfamiliar place, they are insecure and often anxious and eager to prove themselves. This combination may account for the fact that they are so successful in the sciences.

In the past, Britain has thrived because it has been quite an open and tolerant society, as we saw from that Council to help refugees from Nazi-occupied Europe. It is therefore really important to maintain our ability to recruit internationally. To be the best we really do need to recruit from the best. Not everyone shares this view. I recently attended a meeting where Peter Lilley, a very hard-right Brexit-supporting Conservative MP, launched into a tirade about how we were avoiding the problem by importing brains when we should be training young people here. I pointed out that, by recruiting the best people from abroad, you are also providing the best environment for home-grown talent as well. This in turn trains British scientists to be internationally competitive. Thus recruiting the best scientists internationally and training our best young minds is not an either/or thing. Rather, one helps the other. Additionally, migration is not a one-way street. I always encourage young British scientists to go abroad, especially to the United States, to broaden their minds. They learn new techniques and new ways of doing science, and return as better scientists as a result.

To conclude, science has always been an international endeavour. The movement of people has helped the spread of knowledge and accelerated our advancement. Current trends to put up barriers to this exchange will be detrimental to humanity.

[3] Anderson, S. (2014). *The Increasing Importance of Immigrants to Science and Engineering in America.* Arlington: National Foundation for American Policy.

Further Reading

Al-Khalili, J. (2010). *Pathfinders: The Golden Age of Arabic Science*. London: Allen Lane.

Dasgupta, S. (1999). *Jagadis Chandra Bose and the Indian Response to Western Science*. Delhi: Oxford University Press.

Finch, J. (2013). *A Nobel Fellow on Every Floor: A History of the Medical Research Laboratory of Molecular Biology*. London: Icon Books.

Kanigel, R. (1991). *The Man Who Knew Infinity: A Life of the Genius Ramanujan*. London: Abacus.

Miller, A. (2005). *Empire of Stars: Friendship, Obsession and Betrayal in the Quest for Black Holes*. London: Little Brown.

Ramakrishnan, V. (2018). *Gene Machine: The Race to Decipher the Secrets of the Ribosome*. London: Oneworld Publications.

Ronan, C. (1983). *The Cambridge Illustrated History of the World's Science*. Cambridge: Cambridge University Press.

Sime, R. (1996). *Lise Meitner: A Life in Physics*. Berkeley: University of California Press.

Singh, S. (1999). *The Code Book: The Science of Secrecy from Ancient Egypt to Quantum Cryptography*. London: Fourth Estate.

8 Animal Migration

IAIN COUZIN

Ever since I was a kid, I've been fascinated by the type of footage you see of silverside fish schooling near a coral reef and starlings forming these large murmurations in the sky. And what's really remarkable, to me as a biologist, is how little we know about both how and why unrelated organisms like this coordinate these remarkable collective patterns. And yet collective behaviour is not only all around us – it's within us. Our bodies, of course, are a collective of cells. At the Max Planck Institute for Ornithology, we really try to understand the principles of collective behaviour across scales of organisation in biological systems. In this chapter, I'm going to set out some of the latest research on collective animal migration: how organisms come together and integrate their minds to solve very complex problems during migration across the globe.

Insect Mass Migration

I'm going to start with a story about insect mass migration, and I'm going to be talking about a very famous organism, the desert locust. Many of you will have seen footage: when these biblical plagues really are out of control, they are very devastating to the environment. There can be many billions of individuals, some of the largest groups ever found in nature, extending over hundreds of square kilometres. But, in actual fact, when locusts hatch out these little individuals do not have wings. So for the first part of their life as juveniles, they form these flightless swarms. And this is critical because these inevitably precede the flying swarms. These flightless swarms, then, are where we must be targeting for control efforts, as, by the time they take to the wing, it's like a wildfire that's gone out of control.

When I started studying these animals, I noticed that this collective behaviour almost looks cooperative; it's like they are marching in unison together. But why? Why do these insects form swarms?

I do my field research in West Africa, in Mauritania, but this one species of locust, the desert locust, can invade up to one-fifth of the earth's land surface during plague years, and it affects a huge number of people on the planet. These people are often subsistence farmers, so there's hardly any research going into trying to control these pest insects. This is despite the fact that the Food and Agriculture Organization (FAO) estimates that locusts during these plagues, due to the impact on food production, impact the livelihood of one in ten people on the planet.

Now what really got me into action was a wonderful article in a science magazine, where it was written 'Even after 50 years, fighting locusts is more of an art than a science.' And this is very embarrassing to us, especially those of us who come from Europe, because many of us would have locusts at the back of our classrooms as schoolchildren. Many of us who did biology at university would have dissected locusts. But the last time anyone considered what happens when locusts come together was a woman called Peggy Ellis in the early 1950s, and she published her ground-breaking research in the little-known *Anti-locust Bulletin*.

So we wanted to ask some basic fundamental questions: why do insects like locusts exhibit large-scale collective migration? We simply didn't know. And what are the biological processes that underlie this phenomenon, in terms both of how it happens and also why?

In order to answer these questions, I created a sort of particle accelerator for locusts, and I developed software that could track the motion of all of the individuals simultaneously so that we could obtain detailed information about how and why they're moving together in these swarms.

What we discovered was that, if you take all of the different trajectories, you can find what's called an 'order parameter' that tells us whether the locusts are marching clockwise or counter-clockwise around the arena. We have graphs of eight hours' worth of data, which is a typical marching day in the life of these insects. It turns out that upon simply increasing the density, adding more insects into this arena, we see the locusts suddenly adopting counter-clockwise motion and suddenly switching to clockwise, then marching together in what's called an intermittent regime. So just by changing the density of these insects we get a transition from a gas-like state to a driven fluid-like state.

We wanted to understand what's going on, and this reminded us of something that might initially appear to be very different, and that's the behaviour of little particles within magnets. Many of us again have played with magnets and iron filings when we were kids, but if you stop to think why the magnet is magnetic, it's because these little particles in the magnet are all aligned with each other: the collective phenomenon resulting from this is magnetism.

There's lots of very interesting physics about this phenomenon and we wondered whether we can think of the model locusts almost like mobile moving magnetic particles, where individuals interact locally with their neighbours, and like these magnetic particles, they prefer to align with their nearby neighbours. So we could create, in the computer, virtual swarms of insects following these little-magnetic-particle rules to represent the motion of the real animals.

What we found was that this very simple representation was very accurate at capturing the collective properties. If we look around us at the physical world, there's lots of different ways molecules can combine together, yet we find solid, liquid, and gas at the collective level, also plasma if you play around with physics.

Similarly here, we knew from the principles of physics that, despite the details of the individual components, despite us not knowing much about it, we would expect to understand these collective properties regardless. And that indeed was the case.

We had this new theory for how these insects changed from disorder into this ordered collective motion, and it wasn't the individuals themselves changing their behaviour; just changing the density explains everything. But this explains to us how interactions over the scale of 15 cm can scale to swarms that extend over hundreds of square kilometres all moving together, much like, as I said, the magnet having magnetism across the whole structure.

But it didn't explain to us why the locusts did this. And again, I love doing biology because often you get surprises when you do studies of this kind. I was putting the locusts in in the morning and then my colleague,

Jerome Buhl, would take them out in the evening. And he was getting increasingly frustrated with me because I was putting in the wrong number of insects. He was counting them as they came out, and there would be the wrong number. I thought 'Oh goodness, I'm going crazy.' I was really carefully counting, but it was happening again and again, and I thought Jerome was really going to have it in for me.

So I decided to look at the videos myself, and indeed I was counting correctly. The locusts were literally disappearing; they weren't escaping, they were disappearing. And then, when watching these videos, I found frequent examples of them attacking each other, these aggressive inter-actions, but of course these are vegetarian insects . . . or so we thought.

An insect might only suffer some light damage to the abdomen, but, being a desert insect, to lose the haemolymph, the insect equivalent of blood, is very dangerous, and this poor locust is doomed to death. And this gave us an idea that in actual fact perhaps these are not cooperative animals at all; perhaps they are cannibalistic.

So my colleague Sepideh Bazazi conducted experiments whereby she cut a little window into the belly of the locust and found the nerve that gives sensation to the abdomen, then she could cut that nerve so these locusts could no longer feel others biting them from behind. And she did control experiments where she located the nerve, tickled it with a paint-brush, and then sealed it up again. In this case, the locust underwent the same operation, but it still has the nerve, it can still feel.

When we put swarms of nerve-cut and control insects in our arena, we found that the nerve-cut individuals become quiescent. And they are actually statistically identical to individuals that are not interacting at all. We've completely removed their capacity to swarm by preventing them from feeling the biting of others from behind.

Remarkably, these so-called vegetarian insects have actually evolved to be strongly cannibalistic. In these desert environments, where they are really short of liquid, salt, and protein, they turn upon each other. So we have a new mechanism for collective motion that we call a forced march. Far from being cooperative, every individual is trying to eat those ahead and trying to avoid being eaten by those behind. Stop and you risk being cannibalised. So it gave us great insights into how and why these swarms tend to form.

Collective Cognition

Now, if we look elsewhere in the natural world, we also see collective behaviours, but, thankfully perhaps, the animals aren't always trying to eat each other. You can see why in the 1950s, and even into the early 1960s, it was believed there must be telepathy or thought transference among the individuals allowing them to communicate and choreograph their behaviour. But, in fact, we've shown, using computational models, that this is not the case.

What I've become fascinated about is that we know quite a lot about how the individual brain works, how the individual mind works, but we know much less about how social interactions link us together to create collective minds or collective cognition within animal and human groups.

And this beautiful synchronous motion is due to these local interactions among individuals. However, we can't just think and use verbal arguments to understand what's going on. When I was doing my PhD on this topic, we had no data on this phenomenon at all, and so I turned to using computer simulations to understand what's going on inside the mind of the individual. You don't need to know anything about mathematics to understand the concepts of these models; they are very visual and we could represent individuals like fish, but they could also be different types of animals or even cells. The important aspect here is really that these individuals are self-propelling through the environment.

What we found is that there's a strong tendency to avoid getting too close to your neighbours. (For birds in flight it can be fatal to collide.) But equally, being isolated is also dangerous, and you lose the benefits of group living, many of which I'll discuss later on. So there's also a tendency for individuals, for example, to be attracted to other individuals and to align their directions of travel with near neighbours.

It is these local interactions that turn out to be important. There's no global choreographer, there's no one conducting this group telling them where to go; individuals are just following these simple local rules of interaction and that is sufficient to explain many of these remarkable collective properties. These waves of information transfer across the group as individuals turn and respond to those which have turned, which propagates as a fast wave through the structure.

I really wanted to go beyond the modelling approach and try to understand an actual experimental system, to look at these dynamical properties. Lots of theory, often very elegant theory, in animal behaviour has been developed, but many systems are actually dynamic and very few of those theories consider the dynamical properties.

Take, for example, golden shiner fish. These are bred in the billions for live bait in the United States. You can get a thousand of these delivered to your doorstep for $70 including delivery costs, so it's very convenient to work on these animals. Also, despite the fact that they can live in very deep lakes, they're always found at the surface of the water, so it makes the tracking much easier. If we're trying to understand these dynamical properties, we can't just use classical behavioural observation. Instead, we had to develop – over about 10 years – software that can identify and track the individuals, and doesn't lose individuals when they overlap, even though a human observer would. We can now track in real time up to 800 individuals simultaneously.

But why are they forming groups? Well, we wanted to set our fish a computational challenge. Now, these individuals are suffering from a major threat of predation. They have around 92% mortality at this age through predation. So what they will tend to do is to go within the dark regions of the environment, where they're cryptic, where they are hard to see for airborne predators. The fish want to be in the dark regions where they're safe.

We can put a more complex environment in, where there are local dark optima but only one globally best place to be. And, indeed, we found the first experimental evidence that as group size increases the individuals become much better able to solve this problem of finding the optimal place.

We thought this is terrific, we can now look at the accelerations of the individuals with respect to the environment and with respect to each other, because we were thinking along the lines of some very old, very famous work by Francis Galton, a cousin of Charles Darwin, who went to a livestock fair. Galton knew very little about livestock. There was a competition to guess the weight of a dressed ox, and he realised he had no chance, but he also realised almost 800 villagers had already entered the competition, so he calculated the mean and the median of their guesses and was 1 lb off the true weight of the ox.

We thought that perhaps our fish are behaving in the same way. This has been termed the 'wisdom of crowds': each fish must be making a local noisy estimate of the gradient and then, unlike Galton, who could look at all these guesses and add them together using mathematics, our fish might just be adding things together using their fluid-like motion through their social interactions.

We looked at our data to see whether we could find the first evidence of this in nature, in a non-human system. But again, it is great fun doing biology because no, we found no evidence. This led us to a conundrum. If the individuals are not detecting the gradient, how on earth is the group able to do so?

We discovered that natural selection had found an elegant, simple solution. The individuals are measuring the environmental properties exactly where they are, they are not integrating either in time or in space. All they are doing is moving faster in light regions and slower in dark regions. Now this means that, on average, you spend slightly more time in the dark. When they form groups, the interactions among the individuals give rise to what's called an emergent property and a capacity for the group to sense the gradients even though no individual can.

It's like cars on a motorway: because time to collision is the important thing, when cars are moving slowly they tend to be very dense, and that's the same as with our fish. So when they're in the dark they tend to become more dense, they almost solidify within these dark regions, while the rest of the group remains fluid. So the group is computing the gradient – they've evolved to let the group have this property. We've also done evolutionary modelling to show that entirely genetically selfish individuals will evolve into this regime.

Also, our golden shiner fish have now inspired small robots called shiner bots. Each robot is pretty dumb, but by interacting together they can climb gradients, for example of light. And the idea here is that, if we can understand these collective computational properties, we can potentially make robots that are small enough to go inside the body for medical purposes, such as to detect gradients indicative of tumours.

The hallmark of collective behaviour is the fact that an individual changing behaviour can change the behaviour of those that they're socially interacting with, that can in turn change the behaviour of those further on.

Networks of Communication

Now, if you're interested in individual behaviour, of course you can go into the brain and look at the circuits that give rise to individual behaviour. However, when we're looking at the circuits that give rise to collective behaviour, of course these interactions have no physical form. But we've developed techniques that allow us to visualise the invisible. These are quantitative predictions of the probability of you changing behavioural states if I do and vice versa.

We can go in and see that these networks are very complex; there are different weightings, different probability strengths, and different directions. I might be very strongly influenced by you, but you not by me. And so we can then reveal what I find absolutely beautiful: networks of communication. Crucially, not only in a simulation, but in real animal groups, we find predictive networks of communication. We can then ask, what is the relationship between social influence and the structure of this network?

Now, previously, people had assumed that those individuals who are most socially influential must be those who have many connections to others – in the network literature this is called a high degree – and whose neighbours have a low cliquishness – in the network literature this is called a low clustering coefficient. Why is this assumed? This is assumed because people have always thought of the spreading of behaviour as being like an epidemic, like a disease, and there is a very rich body of work in understanding how disease spreads. So they are considering behaviour to be like a fast-spreading disease.

No one has ever been able to ask whether this is actually true for the spreading of behaviours in natural systems, but we can now ask that directly from our data. We can ask it of the experimental data: can we find the first evidence that this is actually the case? Well, no. It turns out that the most socially influential individuals, the so-called super-spreaders of the behavioural contagion, are those with a low degree and whose neighbours have a high clustering coefficient, exactly the opposite of what has always been assumed. But of course we can ask the more interesting question, which is why is this the case?

It comes down to a fundamental difference between disease spreading and social spreading. Social behaviour is an example of what's called a complex contagion, whereas disease is typically a simple contagion.

Let's return to the analogy that I have some infectious disease. The longer I spend in your company, in close proximity to you, the higher the probability, if you're not immune, that I will infect you. However, let's consider behaviour, let's consider a political opinion, say regarding Donald Trump. I may try to convince you that he is a great guy, and I could spend all day trying to convince you, but you may not be convinced – I hope you wouldn't be convinced, in fact. So there's something fundamentally different going on.

Perhaps Donald Trump is too polarising, but if you think of another opinion I may have, I may convince you a little bit, but you don't actually change your mind yet. But perhaps I convince you a little bit, but then later I speak to you and I convince you, and then you talk to someone else and they convince you, and then you talk to the two again, and you're convinced now. In this case, I have influenced you both directly and also indirectly through this loop, and these loops turn out to be very important because we've studied this in fish, in birds, and in humans, and they're all doing this; they're seeking reinforcement from what seem to be independent sources.

Humans do this by knowing who's who, and you'll discount information from close friends and exaggerate the significance of information from individuals that are perceived to be in a different social network. The fish are doing it by just influencing themselves by information coming in from different parts of the visual field. If it comes in from the same part of the visual field it could be correlated, so the fish discount it, whereas if it comes from somewhere else, the fish will attribute significance to the information. Similarly among humans, if the same opinion is encountered beyond your circle of close friends, you'll tend to copy it.

At the beginning I said we can't tell whether it's going to lead to an avalanche or just peter out; suddenly this complex random, stochastic biological system becomes highly predictable. We can predict, before it happens, whether a behavioural change will lead to a cascade propagating an epidemic throughout the group or will peter out, solely by understanding this new relationship between network structure and social contagion.

But it is very difficult to do these experiments because causality is so hard to infer. If an individual changes behaviour, then, with the first individual that responds to that, you can be sure it's responding because of the initiating individual. But then, if you've got another individual responding, is it responding to that initiator or to the second individual or somehow to both? What about the third, the fourth, the fifth, the 100th? Well, it is very difficult to establish what is going on.

We need to have a control. So we have developed, very recently, photorealistic holographic animals, in our case fish, that can swim and interact in the same environment as the real animals. This sounds very much like science fiction, but remarkably this illusion works.

The fish is part of a three-dimensional environment, where it interacts with photorealistic other fish, so there's a real zebra fish at the top and a photorealistic virtual fish below. We project this onto a ball and we track the animal with four cameras, and we can create simple environments. We also created a virtual fish that moves exactly like a real fish. The virtual fish interacts in three-dimensional space, as a hologram, with the real animal.

We can even create what we call the Matrix, where we link together these different holographic worlds. So each animal is in its own hologram, but they are interacting with the data from the others. In this case, they are interacting with each other and the holographic versions of other real individuals. This allows us to study and manipulate more complex group and individual behaviour.

Collective Decision-Making during Migration

In the last part of the lecture I want to talk about leadership and collective decision-making during migration and other behaviours. How do grouping animals make informed decisions? I'm going to start off with some pretty old work first, because it sets the scene for some very recent work.

Organisms are smart; they have goal-oriented behaviour. A certain individual, for example, may be more experienced than others and knows where to go during the migration, whereas others do not know. Does it need to signal to others 'I know what's going on, follow me'? Do other individuals need to be able to recognise who does and who does not have information? Well, we've got no evidence that this is the case. To test

175

this, I created a simple computer model that represents the social interactions and the goal-oriented behaviours. For example, I can colour the knowledgeable individuals in bright colours to distinguish them. But, importantly, in the model, you don't know who agrees with you or disagrees with you, who has information, who does not; the colours are simply for us.

We wanted to ask, can individuals, without signalling, without individual recognition, convey information regarding a migratory route to others? What we found is that one informed individual cannot convey the information to others. It requires a group. Five informed individuals will occupy the frontal edge of the group and the information begins to propagate, and they sort of move in the right direction. Then, if we look at the case with 10 informed individuals, almost spontaneously they start to move very accurately.

Now, we can quantify this, we can measure how effectively they are migrating from 0, which is completely random motion, to 1, which is perfect migratory motion in that preferred direction. We use colours to represent different group sizes. And it turns out that, regardless of group size, as we increase the proportion of informed individuals, we increase the accuracy of migration and then it plateaus off. If we look at group size a little more carefully, let's say we choose 85% accuracy, then, for a small group, like a group of 10 in this example, we need around half of the individuals to have information. But for the same accuracy for a group of 200 individuals, we now need less than 5% of the individuals to have information. We have simulated large migratory flocks and herds and even migratory cells, and it can be an infinitesimal proportion of individuals that know where to go, yet it appears to the observer as if everybody does.

Now, so far we've been talking about cases where everybody was in agreement regarding where to go. However, in many cases within groups there can be disagreement about where to go, what to do, or when to do it. We've also simulated and modelled these types of disagreements. Now remember, nobody knows whether anyone else agrees with them or disagrees with them, you only know your own informational status, yet these groups are very good at voting on the move. They cannot count, they cannot explicitly vote, but the fluid-like nature of the group acts like a collective computational unit. Now, if individuals are very unwilling to give up on their opinions, and they are very strongly opinioned, of course the group can enter a regime where they will split. And so I became very interested in this idea. In fact, we conducted experiments in a football stadium in Germany to actually show that such principles also relate to things like evacuation in human crowds.

There are many examples in humans and in animals where individuals are in disagreement. If, for example, you're on a hiring committee and nobody can agree on a candidate, that's bad for the institution. If individuals within a group can never decide where or when to go, they might be very susceptible to predation or may never migrate at all, which is the worst thing that they could do. And so I got interested in this conflict within groups.

It turns out that when you have this majority versus the minority, then, as I said to you before, the majority will tend to win. But you can then ask yourself, well, what if the minority just becomes more strongly opinioned, intransigent, unwilling to give up on where they want to go? Can that help them get their own way? Well, it turns out that, yes, it can. Being opinioned is a good thing for you because you can manipulate others, but of course those individuals could then become more strongly opinioned to try to counteract this, and this will lead to group fracture.

So we wondered to ourselves, what about if there are individuals in the group who don't know what's going on or they are perfectly well informed but they are unbiased, they don't care whether you go this way or that way? Previously it was always thought that uninformed individuals just add noise, they're not going to do anything interesting. However, this is not the case. Of course, if everybody is uninformed the group can no longer make decisions. However, we discovered that a small proportion of unbiased individuals increases the probability that the group will reach consensus; they act like a social glue that keeps the opposing fractions from just splitting or never seeing eye-to-eye. They increase the speed of reaching consensus, they make decision-making much, much faster, and they make it more accurate. So, far from adding noise, they're doing something much more interesting.

Of course, I'm a biologist; I'm never going to believe this stuff unless we can test it experimentally. So, turning to our beautiful little fish, we wanted to test this theory that uninformed or unbiased individuals should inhibit the influence of a strongly opinionated minority and return control to the majority, namely they should democratise the decision-making process. Working with our fish, we trained individuals to have a preference for either a yellow or a blue target. Yellow turns out to be the crack cocaine of the fish world, they have a much stronger preference for it. It's an innate preference that they have. So they are going to be our strongly opinionated individuals and those trained to prefer blue are our less strongly opinionated individuals.

Now, when we have a minority of strongly opinionated individuals our model predicts that for a small absolute number of uninformed individuals we should see a dramatic change in the group decision, going from a minority-preferred direction to the majority, democratising the decision-making. With the real fish, if there's just minority and majority, and the minority are strongly opinionated, then the minority will get their own way. But, if we put in five or 10 uninformed individuals that don't care where they're going, we return control to the majority; they've democratised the decision-making. They've prevented the strength of opinion from mattering.

Now, this is quite an interesting result because it suggests that diversity of opinion strength, and particularly individuals with low opinion strengths, actually can have a massive impact on decision-making both in groups and in populations. We then turned to another system to try to test these ideas. We put GPS collars on wild baboons in Kenya, and accelerometers on almost all adults within a wild baboon troop. Now, of course, for ethical reasons, we did not tag juveniles. None of the animals were harmed, of course.

We found tens of thousands of events where some individuals move one way and others move in a different direction, but, like our fish, they don't want to lose the benefits of group living, so one subset has to give up. And, really remarkably, we found that, unlike what you often hear in natural history programmes, it doesn't matter a jot whether you're a dominant male going one way and a subordinate female going the other, its 50:50 which one will be followed.

Social dominance, which of course is all about aggression, doesn't matter and yet, when we look at our theory of how these groups make decisions, we found that the baboons behaved exactly as I described to you before, really remarkably so. Now, it could be that there are uninformed individuals in the group and that democratises the process, or it could be that the dominant individual wants the collective to make decisions about where to go because when they get there – I've watched these animals in the field – the dominant individual can completely monopolise the resource. So it may still be in the interest of the dominant male individual to let the group exhibit collective intelligence; we don't know yet.

But of course these animals are not like fish in a barren environment, they are living in a complex physical environment. We have generated a three-dimensional point cloud taken from a drone in Kenya. This drone automatically maps the three-dimensional vegetation structure down to around a spatial accuracy of 3 cm. We can get great detail now of the complex physical environment in which these animals have evolved. You can even see the single tyre tracks in the road.

Using all this data, we found a great surprise. The baboons are not interacting with each other as points in space, even in dense environments. What they are doing is they are following the path, the footprints, of where others have been within the last five minutes or so. Probably a little bit like ants, they are able to smell where others have been in the last five minutes and tend to follow the odour of others. Putting the habitat and the social interactions together therefore allowed us to understand how these animals behave.

I think we're at a critical point in the study of animal behaviour and migration particularly, where we can put new, small, cheap, lightweight devices onto a much wider range of animals to get quantitative data.

Recently, in a project led by my colleague Martin Wikelski, we've put a new device onto the International Space Station. The first went up in October, the second piece went up last month, and the cosmonauts will do a five-hour spacewalk to attach a big receiver to the International Space Station. Now the importance of this is that we can use these new, small, cheap, lightweight designed transmitters and we don't have to recover the animal. Wherever the animal is on the globe, except for the Poles, we can recover the data via the Space Station. And so it's a really exciting time to do these studies.

We can actually have data on every flap that these birds exhibit over their entire lifetimes, and we can look to see how they're buffeted by the winds, and that allows us to reconstruct the detailed physical air environment through which they move. Martin wants to create a network of animals across the planet that gives us data from regions we can't possibly access. We now have real data – from GPS and accelerometers – of storks riding thermals. And there are some very interesting data that we can now use to understand the different roles that individuals play.

Lastly, I just want to mention that we also work in the field using drones, where we can use a drone to track these animals. And using the latest state-of-the-art deep-learning methodologies, we can now track these animals, including where they're looking, as they are undergoing migrations. Using these different types of new technologies, we're getting a completely new window into the lives and migration behaviour of wild animals.

Index

Index